T0132812

Eyewitness:

Evolution of the Atmospheric Sciences

Eyewitness:

Evolution of the Atmospheric Sciences

By Robert G. Fleagle
University of Washington
Seattle, Washington

Historical Monograph Series
John S. Perry, Historical Monograph Editor
Boston
American Meteorological Society
2001

ISBN 1-878220-39-X

Library of Congress Control Number: 2001135211

Published by the American Meteorological Society
45 Beacon Street, Boston, MA 02108

Ronald D. McPherson, Executive Director
Keith L. Seitter, Deputy Executive Director
Kenneth Heideman, Director of Publications
Andrea Schein, Copy Editor

Printed in the United States of America
by Sheridan Books

Contents

Preface

My professional career has extended over a period of remarkable technological developments that have led to dramatic advances in the atmospheric sciences. For many of the seminal advances I have had a seat directly behind home plate, for others a bleacher seat, while for some I could only hear the crack of the bat or occasionally glimpse a ball arcing above the grandstand. The following account describes with broad strokes what occurred over the more than half a century of my career and attempts to explain how and why some of the critical events occurred. I have relied on personal memory, on letters and personal documents retained in my files, on published papers and books, and on selected reports of institutions, agencies, and committees. Several significant events were remembered only after reference to my files, evidence, if any were needed, that personal memories can be selective and treacherous. Others who experienced the events described here may have perspectives that differ from mine. Colleagues identified under the acknowledgments have helped by reading an earlier draft and correcting errors and identifying points needing clarification or elaboration.

Changes of the past half century have been driven by new technologies that now provide data unimaginable at the beginning of the period, while equally unimaginable capabilities have evolved for using the data productively and for archiving it efficiently. It is now possible to examine a myriad of natural phenomena that were formerly inaccessible, to discover new phenomena, to introduce new applications, and to reveal potential environmental threats arising naturally or resulting from human actions that were formerly unknown. The numbers of atmospheric scientists and their specialties have proliferated, and new institutions have been created and have evolved over the decades. The level of scientific and technical sophistication of the field has risen markedly. Several dimensions are required to describe evolution of the atmospheric sciences. A single indicator is revealed by noting that volume 1 of the *Journal of Meteorology* (1944), published by the American Meteorological Society (AMS), contains 116 pages, while volume 57 of the *Journal of the Atmospheric Sciences* (2000) contains 4048 pages.[1]

---

[1] The change of name occurred with volume 19 in 1962.

And nine additional journals are now published by the AMS that did not exist in 1944.

Imagine an observer watching as a passenger train slowly leaves the station platform. The first car is clearly seen to contain only a few passengers, each of whom the observer knows personally or by reputation. In the second car there are more passengers, and it passes more quickly. As subsequent cars pass still more quickly, individuals are more numerous and harder and harder to identify, and many are unknown to the observer. Finally, the crowded cars pass so rapidly and there are so many passengers that only an occasional face can be identified peering from a passing window. At the same time, as the cars pass some of the train's basic characteristics—its length, relation of coach and pullman sections, location of dining and observation cars—can be understood more completely. That has been my experience as the decades have passed. My eyewitness observations of events become less detailed, but perhaps more perceptive and comprehensive, as the memoir unfolds.

The observations that follow can provide an extensive supplement to my interview recorded in 1989 by Earl Droessler as part of the historical interview project of the American Meteorological Society and the University Corporation for Atmospheric Research. They may provide a side light as historians peruse the twentieth-century record of the atmospheric sciences and examine how science and public affairs interacted over that period. And perhaps former students and colleagues may be interested in my impressions of events that have led to the scientific and professional activities that are shaping or have shaped their careers.

Although several chapters focus on my particular experiences as a graduate student at New York University (NYU) and a faculty member of the University of Washington (UW), they concern topics that may have resonances in the experiences of many of my contemporaries at other universities and therefore may have general relevance to this account of the evolution of the atmospheric sciences. Other chapters dealing with the institutions that I encountered along the way describe events that also were experienced by many of my colleagues, perhaps in different ways.

The terms *meteorology* and *atmospheric sciences* are both used here to denote the science of the atmospheres of the earth and planets; the distinction between the two terms is largely historical. Each embraces motions on all scales, atmospheric radiation, climatology, cloud and aerosol physics, atmospheric chemistry, and atmospheric electricity.[2] The science of the atmosphere overlaps and interleaves with other geophysical and astrophysical sciences, and with biological sciences, so there is no attempt to draw sharp boundaries.

An underlying theme throughout concerns the influence, sometimes stimulating and sometimes constraining, of politics on development of new institutions and on scientific achievement. Political issues range from those of strategic geopolitics, to national party politics, to parochial issues involving federal agencies, departments, and congressional committees, and to equally parochial issues among science organizations or individual universities and centers.

The following chapters are arranged in rough order of major events in my career, although detailed topics are often carried beyond the chronological or institutional bounds of a single chapter; and there is considerable interleaving of elements. Early focus on science shifts gradually toward policy issues as the memoir unfolds. The first chapter concerns the period of my graduate study at NYU, the second the early development of the atmospheric sciences program at the UW. Later chapters are concerned with political events affecting the UW and the atmospheric sciences generally, and with steps in evolution of the field as they were affected by major research programs and by government agencies, the National Academy of Sciences, the University Corporation for Atmospheric Research, the American Meteorological Society, Congressional committees, and local environmental groups with which I have worked in the Pacific Northwest.

---

[2] Physics of the ionosphere and magnetosphere is not included.

CHAPTER 1

Graduate Study

When I entered meteorology during the Second World War the field was in the process of transformation from a largely descriptive field to one based more solidly on scientific principles.[3] The first meteorology program in this country to offer doctoral degrees had been initiated in 1928 at the Massachusetts Institute of Technology (MIT), an initiative stimulated by the fact that meteorology was more advanced in Europe than in this country. Carl-Gustav Rossby, a Swedish meteorologist trained at the University of Bergen, was appointed to lead the program. In 1930 the MIT Meteorology Department was established and Rossby became its chairman; in the following decade research conducted under Rossby's leadership included projects in dynamic and synoptic meteorology, boundary layer turbulence, atmosphere and ocean observations, air–sea interaction, and instrument development.[4] Six graduate students in the program earned Doctor of Science degrees; several others who did not complete doctoral degrees went on to distinguished careers, notably Henry Houghton, Jerome Namias, and Athelstan Spilhaus. In 1939 and 1940, at Rossby's initiative and with military support, similar programs were established at New York University, University of Chicago, University of California at Los Angeles (UCLA), and California Institute of Technology (Cal Tech). During the next four years these five universities provided nine months of graduate training for several thousand students who had earned bachelor degrees in the physical sciences, engineering, or mathematics and who then served in the weather services of the Army Air Forces and the Navy. Later, undergraduate stu-

---

[3] I became aware of meteorology as a scientific field and a profession in 1939 when, as an undergraduate physics student at The Johns Hopkins University, I discovered W.J. Humphreys's *Physics of the Air* (Humphreys 1929). In 1941 or 1942 I read George Stewart's *Storm* (Stewart 1941) and this was a spur to my applying for meteorological training.

[4] Rossby's career and circumstances of his appointment at MIT have been described by Norman A. Phillips (Phillips 1998) and by Frederick Nebeker (Nebeker 1995).

dents were selected for equivalent training; and others received training at a military school at Grand Rapids, Michigan.

The field of meteorology in the early 1940s was well defined by the curricula of the five departments offering graduate degrees—dynamic meteorology, physical meteorology, synoptic meteorology, climatology, instruments and observations, analysis and forecasting, and occasional lectures in special topics. The focus of the courses was on 6–12-h forecasting for aircraft and naval operations.

Meteorology was data limited and data poor to a degree hard to appreciate today. Sea level pressure fields and upper-air flow fields were plotted and analyzed routinely, by hand of course; pressure, temperature, and moisture fields were analyzed over the United States using sparse radiosonde data; pilot balloon observations provided limited definition of wind fields, while more detailed operational or research data could be obtained only by special efforts using relatively primitive in situ instruments.

I arrived at NYU on December 2, 1942, the same day, I found later, that Enrico Fermi achieved the first controlled nuclear chain reaction at the University of Chicago. Enrolled in the "5th war course" were about 400 aviation cadets, including me, fewer than 10 navy junior officers, and a few civilians. The faculty at NYU at that time included Raymond B. Montgomery and Hans A. Panofsky in Dynamic and Physical Meteorology; Gardner Emmons, James E. Miller, and Robert Culnan in Synoptic Meteorology; and Yale Mintz in Climatology. Athelstan Spilhaus, chairman of the department, had been commissioned in the Army Air Forces (AAF) but gave a few lectures on instrumentation and map projections. Acting Chairman Emmons was a master synoptic analyst, and our courses emphasized surface map analysis, with less attention given to analysis of soundings and of upper-air charts. Several students from earlier classes were assigned as laboratory instructors; among this group were Homer Mantis and Herschel Slater, who were my instructors, Joshua Holland, Conrad Mook, Joanne Gerould (later Simpson), Fred Decker, and Silvio Simplicio, whose specialty was lecturing aggressively on the heavy responsibilities of forecasting for military flights. The program inevitably was impersonal, heavy on direction and procedure, and light on clarifying explanations. Textbooks were supplemented by hundreds of pages of mimeographed lecture notes and class instructions. There were frequent quizzes, and students were rated on everything they turned in. Military inspections and drill occupied an hour or more per day, and we cadets were not allowed to forget that we were subject to military discipline.

We got a heavy dose of air mass and frontal analysis and of Petterssen's forecasting methods.[5] We were indoctrinated with the belief that a good analyst should be able to find a front to account for any area of nonconvective cloud or precipitation. We accepted the view that storms were manifestations of the instability of fronts, though we realized that this had not been proven.[6] A glimpse of better things appeared briefly when Gilbert Hunt, a member of the class, gave a special report on Lewis F. Richardson's attempt at numerical forecasting. This was stimulating to me and closer to what I had hoped to encounter when I applied for the program, but it was unconnected to our courses and therefore mystifying. We heard nothing of possible future development of electronic computers that could be used to integrate the hydrodynamic equations. We learned of Sir Gilbert Walker's fascinating correlations linking changes of sea level pressure in the equatorial western Pacific to weather in the southeast United States and other distant sites; this stimulated thought and imagination, but no one seemed to know what to do with the information. It was not until the 1970s and later that this conundrum began to be unraveled.

There was some attention paid to analysis of upper-air flow and analysis of isentropic charts, to preparation of cross sections, and to forecasting for aircraft flights. Near the end of the course we were introduced to single-station forecasting. This was based on the Rossby wave equation and calculation of constant absolute vorticity trajectories, using wind velocities at heights of 10 000 ft at a single station to determine the location of downstream or upstream troughs or ridges. It was an interesting concept, but I do not recall that successful single-station forecasts were ever demonstrated.

Members of my class who later distinguished themselves in other fields included Kenneth Arrow (Nobel laureate in economics), William E. Gordon (ionospheric physicist), and Gilbert Hunt (applied mathe-

---

[5] A personalized account of Petterssen's career is provided in Petterssen (2001).

[6] These ideas came to us from the revered but largely unread bible of meteorology, Bjerknes et al. (1933).

matician).[7] More than 300 cadets completed the course in September 1943 and were commissioned as second lieutenants; most were assigned to weather forecasting duty at airfields.

For the following nine months I was assigned as a laboratory instructor for the seventh war course. By this time the pool of applicants with bachelors degrees had been depleted, and the course was opened to undergraduates with outstanding records. My laboratory group included a remarkable number who later distinguished themselves in the atmospheric sciences, notably David Atlas, Louis Battan, Myron Ligda, and William Reifsnyder.[8] During this time I had opportunities for closer contacts with faculty and discussions with fellow lab instructors, resulting in rapidly expanding my understanding of meteorology. We began to recognize that the concepts of fronts learned from textbooks fell short of accounting for observed variations in storm systems, and soundings and upper-air charts received increased attention. We began to recognize that frontogenesis often occurred with storm development and to doubt that instability of frontal surfaces was the cause of cyclogenesis. We heard more about waves in the westerlies and conservation of vorticity, but the connection to surface frontal structure remained obscure. I think it was at this time that forecasting for periods longer than a single aircraft flight led to interest in index cycles and blocking high pressure systems. I had the impression that the civilian faculty was only a little ahead of us in absorbing some of these concepts.

Hans (Hap) Panofsky was participating in a program of observations of stratospheric ozone using a Dobson spectrophotometer; the objective was to relate variations in ozone in the vertical column extending to the top of the atmosphere to surface weather. I learned to make the Dobson observations, an opportunity that helped to enlarge my ap-

---

[7] We were assigned to seats alphabetically in the large lecture room in Gould Hall, so Arrow sat in the front row, Gordon sat next to me to my right in about the fourth row, and Hunt farther down the row.

[8] Atlas has had a long career as a productive physical meteorologist; he is a member of the National Academy of Engineering, and a former president of the AMS. Battan, who spent most of his career at the University of Arizona, was a leader in cloud physics and a former president of the AMS. Ligda, who died prematurely, was an early leader in meteorological remote sensing at the Stanford Research Institute. Reifsnyder has been a faculty member at Yale University and a leader in forest meteorology and climatology.

preciation of the potential scope and challenge of the atmospheric sciences. Several of us were involved also in efforts to develop, launch, and track constant-level balloons, efforts that were entertaining but largely unsuccessful.

I undertook a research study of nocturnal thunderstorms in the Midwest with the hope that it might lead to a master's thesis. After reviewing data on the distribution, frequency, and severity of these storms, I tried to determine whether they might be initiated and driven by enhanced convective instability resulting from infrared radiation to space from the tops of clouds. I worked on radiation calculations but was not able to explain the timing or geographic distribution of the storms. So my initial research project came to naught. Many years later midwestern nocturnal thunderstorms were substantially explained as a consequence of ageostrophic flow and regional convergence in the upper boundary layer resulting from the diurnal cycle of temperature in the boundary layer driven by downslope flow over the terrain east of the Rocky Mountains (Buajitti and Blackadar 1957; Holton 1967; Bonner and Paegle 1970). After abandoning the study of nocturnal thunderstorms, I began a study of the effect of measured horizontal velocity divergence on temperature lapse rate that I used later to satisfy the thesis requirement for the degree of master of science in meteorology.

New York University played a major role in the multiuniversity Northern Hemisphere Project, whose objective was to produce daily surface weather maps for the 40-year period preceding the war using all available data.[9] Efforts were made to generate a dataset as complete, correct, and homogeneous as possible; but very large gaps in data remained. The project employed teams of data processors, analysts, and a clerical staff that included my wife, Marianne, and wives of other NYU faculty and students. The completed maps were intended as the database for forecasts by what was called "the analogue method." This entailed finding a map that matched the current weather map most closely, and then using the subsequent series of maps as forecasts. The objective proved unattainable. After the war the maps were published and distributed, but had only limited use.

At the end of the nine-month course in June 1944 I was one of a

---

[9] Initially the California Institute of Technology participated in the project, but was replaced later by Chicago and UCLA.

group assigned to visit airfield weather stations to instruct forecasters in new concepts and techniques in weather analysis and forecasting. We first spent a week or two at the University of Chicago where we gained some experience with calculating constant vorticity trajectories and concentrated especially on analyzing constant-level upper-air charts (5000 and 10 000 ft) with the objective of forecasting summer convective storms. This represented an important advance from the heavy emphasis on frontal analysis that had been my experience until then. I was then sent to Wright–Patterson Airfield at Dayton, Ohio, as one of a three-man team assigned to the Midwest region. We traveled by military aircraft and spent 3 or 4 days at each station giving lectures and conducting workshops. Sometimes I felt that it was a case of the blind leading the blind.

In 1944 Rossby's vision and leadership as president of the AMS were largely responsible for reorganizing the Society under a new constitution, launching the *Journal of Meteorology*, and appointing Kenneth Spengler as full-time executive secretary (later executive director). I was greatly stimulated by these events and welcomed the prospect of membership in an organization with high scientific standards and an expanding program. Spengler's appointment was especially notable; over the years he became the indispensable man, providing quiet, insightful guidance to all components of the AMS, developing its programs of meetings and publications, providing tutelage to succeeding councilors and presidents, and nourishing its growth and its financial stability.

In the fall of 1944 Research Weather Stations were established at several of the universities that had trained forecasters, and I was assigned to the station at NYU. Harry Wexler, as chief scientist for the AAF Weather Service, headed the multiuniversity program. Sam Solot was the first director at NYU; and the staff included Homer Mantis, Woodrow Dickey, William Turkel, L.G. Engle, Vernon J. Hansen, me, and perhaps someone I have forgotten.[10] Later, Solot was reassigned, and Mantis became director. Research projects on 24-hour forecasting

---

[10] Solot was a Weather Bureau research meteorologist before and after the war. Mantis's career has been spent largely as a member of the Physics Department at the University of Minnesota. Dickey became a Weather Bureau research meteorologist and forecaster. The other staff members, so far as I know, did not continue in meteorology after the war.

of surface temperature and calculation of synoptic-scale vertical velocity were pursued with guidance by the department faculty. The temperature forecasting project, led by James E. (Jim) Miller, was based on methods to estimate trajectories of boundary layer air parcels together with physical–statistical methods to estimate temperature changes experienced by the parcels as they moved along the trajectories. The vertical velocity project, led by Hans Panofsky, involved calculating vertical velocities at grid points by the kinematic and adiabatic methods and comparing the results. It required careful analysis of wind and temperature fields on upper-air charts. A number of different techniques were used for each method, and in 1945 four reports were written and submitted to the Air Weather Service for limited distribution. In 1945 I presented my first paper on this work at a meeting of the American Geophysical Union. At the end of the paper Jerome Namias, already recognized as a major figure, stood up to comment. To my great relief he complimented me on what he called a new and important contribution. That paper was published in the *Transactions of the American Geophysical Union* (AGU), my first in a refereed journal (Fleagle 1945).

To report on the work on vertical velocity calculations I visited MIT where I met for the first time faculty member James M. Austin and William H. Klein, Robert M. White, and Edward Dolezel, who were AAF officers assigned to the MIT Research Weather Station.[11]

In July 1945 I was assigned to forecasting duty at the weather office on Guam, with orders to report about the middle of August. However, before the date of my departure the bombings of Hiroshima and Nagasaki occurred, my orders were cancelled, and I continued research at NYU until I was discharged from the Army Air Forces in April 1946.

In early 1946 Carl Rossby visited NYU to discuss our vertical velocity work and to tell us about research at the University of Chicago. As I remember, it was on that occasion that I first heard about the con-

---

[11] Klein has been a research meteorologist in the National Weather Service and in the private sector. White has been an especially distinguished science administrator, having been chief of the Weather Bureau, administrator of ESSA and of NOAA, president of the AMS and of the University Corporation for Amospheric Research, and president of the National Academy of Engineering. So far as I know, Dolezel did not continue in the atmospheric sciences after the war. Austin's subsequent professional career was spent entirely as an MIT faculty member.

cept and consequences of group velocity of waves in the westerlies. I wrote to Rossby inquiring about applying to his department for further graduate work and about pursuing my work on vertical velocity as a Ph.D. thesis. He replied encouraging me on both questions. At about the same time NYU announced a new Ph.D. program to be administered jointly by the physics and meteorology departments, probably in anticipation of development of atomic energy and consequent need for closer linking of the two fields. The combination of physics and meteorology was highly attractive to me, and it was easier to remain at NYU. So I enrolled in the physics–meteorology program, as did Homer Mantis.[12] The degree required passing qualifying examinations in general physics, atomic physics, general meteorology, and dynamic meteorology, and a thesis. Subsequently, with the appointment of Bernhard Haurwitz as chairman of the department, a Ph.D. degree in meteorology was approved; and later students enrolled in this program. Included were the following, with whom I was closely associated for most of the period from 1946 to 1948: Jerome Spar, Alfred Blackadar, Willard Pierson, Julius London, Joseph Smagorinsky, Morton Barad, Franklin Badgley, Frank Gifford, Ben Davidson, Thomas Gleeson, Edwin Fisher, Emil Kohler, Harold Stolov, and Louis Berkofsky. Philip Thompson, who was a member of the Meteorology Project at the Institute for Advanced Study (IAS) at Princeton, was a visitor and may have been a graduate student for a time, but completed his degree at MIT.

Discussion of the potentialities of numerical forecasting had begun in 1945 among a group that included John von Neumann [mathematician at the Institute for Advanced Study (IAS)], Rossby, Francis Reichelderfer (chief of the Weather Bureau), Vladimir Zworykin (RCA engineer), and Harry Wexler (who became director of research for the Weather Bureau). These discussions led to establishment of the IAS Meteorology Project on July 1, 1946.[13] Initial tasks were to test numerical models that could be developed to run on computers not yet available, but that could be reasonably anticipated within 2 or 3 years. Hans

---

[12] Adoption of the G.I. Bill immediately after the war made possible expanded graduate programs in all fields of science, social science, and humanities. It provided training for the generation that became the leaders in the evolution of the atmospheric sciences in the United States.

[13] Establishment of the Meteorology Project is described in considerable detail in Nebeker (1995).

Panofsky was enlisted as a consultant with responsibilities for developing methods of objective map analysis and for comparing vertical velocities produced by numerical models with those computed by the methods we were using at NYU. On one occasion, probably in the spring of 1947, with Panofsky I attended a project meeting at Princeton that I remember especially because it afforded my only sight of Albert Einstein. Late one afternoon someone called attention to Einstein walking along a path crossing the lawn, probably on his way home from the Institute. The image of his distinctive figure, seen from the window of the meeting room, has remained clear in my mind ever since. Through Panofsky's participation in the Meteorology Project I gained a peripheral view of the initial steps toward numerical prediction.

From 1946 to 1948 I took courses in Advanced Calculus, Methods of Mathematical Physics (taught by Morton Hammermesh), Laboratory Physics (Yardley Beers), Theoretical Mechanics (Fritz Reiche), Atomic Physics, Atmospheric Oscillations (Bernhard Haurwitz), Oceanography (Panofsky), and Theoretical Climatology (Panofsky). This last was a new course focused on radiational effects on climate and especially on the Milankovich theory of climate change. I believe that at that time Panofsky was virtually alone among U.S. meteorologists in taking the Milankovitch theory seriously as accounting for ice ages. In the spring of 1947 I passed three of the qualifying examinations, but was required to repeat the exam in atomic physics a year later. I also began to focus on my thesis under Panofsky's supervision.

Publication in 1947 of Jule Charney's paper on baroclinic instability represented a major and dramatic scientific advance; it provided a basis for understanding the growth of waves in the westerlies, ended confusion and debate over instabilities of frontal surfaces as the cause of cyclogenesis, and opened a rich field of theoretical research. I was greatly stimulated by Charney's paper, though I found the mathematics formidable and somewhat obscure. I found Eric Eady's work, carried out independently of Charney's (though published more than a year later), more fully accessible. These two papers provided the theoretical basis for the work that I was doing in describing the fields of vertical motion associated with midlatitude cyclones (Charney 1947; Eady 1949).

I was also especially stimulated by Woodrow Jacobs's paper showing cyclogenesis highly concentrated over the Gulf Stream and the Kuroshio Current (Jacobs 1942), a continuing influence that over later years tilted my research interests toward problems of air–sea interaction.

My thesis required determination of fields of vertical velocity of as high quality as possible in three selected weather situations. To obtain the necessary upper-air data, I spent several days at the Weather Bureau in Washington, D.C., copying data from original radiosonde records and recalculating pressures at standard levels. The small Bureau buildings at 24th and M St., NW, held the administrative offices, data archives, forecasting center, long-range section, and probably other components I have overlooked.

At that time the Weather Bureau was being challenged for national primacy in meteorological research by the Geophysics Research Directorate of the Air Force and the Office of Naval Research. My experience had been entirely within the Army Air Forces Weather Service, and I was glad to have this first contact with the Weather Bureau. Tensions were felt and sometimes expressed between those proud of the standards and traditions of the Bureau and those stimulated by the new technologies and new opportunities offered through the larger budgets of the military agencies. This issue appeared in various guises over later years, sometimes inhibiting but more often stimulating progress.

My thesis was submitted in May 1948 and resulted in two publications in the *Journal of Meteorology*, one that described the fields of three-dimensional motion in a vertical cross section through a cyclone–anticyclone system derived as a composite from the three selected situations, and a second paper that evaluated and described the processes resulting in pressure changes occurring in these situations (Fleagle 1947, 1948). As time for the defense of the thesis approached, I realized with apprehension that my committee would be considering awarding the first Ph.D. degree in physics–meteorology. As I remember, the committee consisted of Haurwitz (chairman), Joseph Boyce (chairman of physics), Panofsky, Serge Korff (cosmic ray physicist), and Karl Friedrichs (Courant Institute). I was especially apprehensive about Friedrichs, whose reputation as an eminent mathematician was well known but whom I had not met personally. He proved to be merciful and the exam a pleasant formality.

When I began to look for an academic job in early 1948 openings in meteorology were announced at the UW where a new department had been established in 1947 with Phil Church as executive officer (later chairman), and at the University of Florida, which was interested in appointing someone to develop a new department at Gainesville. Hamilton College announced a position in physics that indicated interest in developing courses in meteorology. The UW position seemed most attractive to me, but I asked others for their advice. The meteorology faculty favored the UW, and Joseph Boyce recommended a state university over an undergraduate college. I happened to accompany Athelstan Spilhaus on a train trip returning from Woods Hole Oceanographic Institution and asked for his advice. His recommendation was to choose the position of largest Coriolis parameter. My positive feeling about the UW position was based in part on having heard Phil Church give a paper at the winter meeting of the AMS in New York the previous January. His paper discussed meanders of the Gulf Stream and used data and concepts that were challenging and new to me. I was also impressed by the fact that Church had reached New York by bus during a railroad strike that had prevented nearly everyone outside the New York area from attending the meeting.

During the spring Phil Church made a recruiting trip and stopped at NYU to meet me. We had a chance to talk about my interest in an academic career and his plans for the Department of Meteorology and Climatology. The program was to be developed gradually and to be modeled on programs of the five departments that had trained meteorologists during the war. An undergraduate curriculum had been approved by the university, and the department was authorized to award B.S. and M.S. degrees. A week or so later I received a telegram offering me an assistant professorship at a 9-month salary of $3,807. I replied by telegram accepting the appointment.

Late in the summer I learned from an article in the *New Republic* that the University of Washington had been accused by a committee of the state legislature (Canwell Committee) of harboring Communists, and that the university administration seemed to be supporting the committee's efforts to investigate the faculty and to have several members dismissed. I was dismayed at the prospect of joining a university that would not defend its faculty against such charges. I felt that the stature of the university had been diminished greatly. It was too late to reconsider my acceptance of the appointment, but I reached Seattle at the beginning of September with reduced enthusiasm and with some apprehension as to my future career. It is likely that if I had known of these events earlier, I would not have accepted the appointment.

CHAPTER 2

Early Years at the University of Washington

However, the atmosphere within the department was positive and reassuring. Phil Church and his wife, Laurie (Loretta), welcomed my family warmly and eased our first weeks greatly. Phil had won approval for increasing my annual salary by $99, a surprise to me; and he arranged to put me on a research payroll from September 1 to 15, when my regular appointment began. The objective of the project was to understand local differences in surface temperature in Alaska; I continued with the project during 1948 and 1949 and published several papers on that work.

Over the years Phil Church's generosity of spirit and concern for the welfare and success of individual faculty infused the department with qualities vital to its development. Visitors and new faculty often have remarked on the friendly, helpful spirit of the department. Phil's influence in creating this intangible factor has been an important element in the achievements of students and faculty and the standing achieved by the department in later years.

The department had been assigned space in the new Thomson Hall consisting of four offices, two classrooms, and storage space. The standard teaching schedule at the UW was 10 credit hours per week. Initially there were no graduate students in the department. Most students were poorly prepared in physics and mathematics, and academic standards of the university as a whole seemed low. I concentrated on preparation for my two daily classes, and directed special efforts at strengthening students' backgrounds in physics and their capabilities in writing accurately and clearly. I taught courses for juniors and seniors in physical and dynamic meteorology, Phil Church taught physical and regional climatology, and William Schallert taught synoptic meteorology and analysis and forecasting. Each of us took our turns at teaching the survey course for lower division students. In 1950 E. Paul McClain joined the department on a visiting basis, replacing Schallert who went to UCLA for graduate study. Also in 1950 I became an associate editor of the *Journal of Meteorology*; this gave me added opportunity and incentive to keep abreast of research going on at other universities and government laboratories.

The Research Society, a group of faculty active in research, met weekly in the old faculty club (the Hoo-Hoo House, a legacy from the Alaska–Yukon Exposition of 1909) to listen to and discuss a report from one of its members. Phil Church introduced me to the group, and I gave a report on my current research and became a member. The Research Society provided an opportunity to learn about research in other fields and to meet faculty from other departments. Paradoxically, as the UW became more active in research the Research Society withered and, as I recall, did not survive the mid-1950s. In retrospect, this can be recognized as resulting from the growth of federal support for university research with resulting shift of faculty allegiance from local to national.[14] It may also have reflected loss of faculty morale due to the political tensions of the time. Both factors contributed to the erosion of collegiality that has characterized academic life since World War II.

The best undergraduate meteorology student in those early years was Dean Staley, who graduated in 1950 and at my urging went to UCLA for graduate study. After completing the M.S. degree there, he returned to the UW to become our first Ph.D. student. Several other students who completed requirements for the B.S. degree in those first years entered our master's degree program. I recall especially Edwin Danielsen arriving in my office in the spring of 1951 to discuss graduate study; his personality and artistic talent contributed much to the department over the following decade. Danielsen's imaginative painting of a vertical cross section through the geosphere graced the entrance to our building for many years until it was stolen in 1969 during the move to our present building.

From 1944 to 1946 Phil Church had directed the meteorological monitoring program at the Hanford Works in eastern Washington, where plutonium was produced for nuclear weapons development and production; after returning to the university he continued as a consultant to that program. At Hanford a 400-ft instrumented tower provided continuous records of temperature and wind velocity at 50-ft intervals;

---

[14] However, two smaller more exclusive research groups, the Catalysts and the Clankers, have served some of the same functions as the Research Society. The Clankers, to which I belonged, ceased meeting in the 1970s, but the Catalysts continue.

these data provided records from which boundary layer diffusion and transport of radioactive particles and gases could be assessed. At the UW Phil planned for the new department to focus on study of boundary layer processes; the first step had been to build a 50-m instrumented tower on the university golf course between Pacific Street and the Montlake cut. Data were recorded from eight levels distributed logarithmically with height. Beginning in 1949 an Atomic Energy Commission (AEC) contract provided research funds for instrumentation and for support of about five graduate students. Frank Badgley, who completed his Ph.D. at NYU in 1950, was appointed as research assistant professor with special responsibility for the AEC research program as well as for teaching; a year later he joined the academic faculty. Over the years his research has included boundary layer turbulence, analysis of the heat budget of the Arctic Ocean, and turbulence in aircraft contrails. He led the department's research group that measured ocean surface fluxes in the International Indian Ocean Expedition. And he has served in many administrative roles including associate director of the Quaternary Research Center from 1972 to 1977 and chairman of the department from 1977 to 1982.

In 1950 graduate courses were introduced in dynamic meteorology and atmospheric turbulence and diffusion; students were required to take applied mathematics and encouraged to take other courses in physics and oceanography. Credit also could be earned in studies of special topics. The master's degree required passing a written qualifying exam and a general exam, initially written but later oral and limited to the topic of the thesis. Utilizing data from the meteorological tower, M.S. degrees were earned by William Parrott (1951, F),[15] James Fuquay (1952, Ba), Paul Davis (1952, F), Richard Hubley (1952, F), and John Scuderi (1952, F) and later by Norman Wagner (1955, Ba) and Harold Twedten (1955, Ba). Master's degrees in climatology were earned by Arnold Court (1949, C); (based on work done earlier in the Geography Department), Thomas Stephens (1952, C), and Charles Cushman (1953, C).

To link our research with related work at Hanford, Phil Church arranged for Frank Badgley and me to obtain security clearances and

---

[15] Each student's thesis supervisor is designated by an initial: Church (C), Fleagle (F), Badgley (Ba), Buettner (Bu), Reed (R), Businger (Bs), Hobbs (Ho), Holton (Hn), Wallace (W), Leovy (L).

to contribute to his work there. Frank's association with Hanford continued throughout his career at the UW and for several more years. Morton Barad, formerly a fellow graduate student of Frank's and mine at NYU, became head of the Hanford group in 1951 and was appointed as a visiting faculty member in our department. In the early 1950s we traveled to Hanford on several occasions. I remember especially attending a discussion of a Hanford field experiment, "Greenglow" I believe, in which fluorescent particles were released and traced in order to test diffusion models. I was puzzled when Dr. Herbert Parker, the senior official responsible for health protection at Hanford, expressed his anger and frustration over unstated and presumably classified aspects of the program that he evidently felt were irresponsible. Following release of the data in 1986, it now appears that the diffusing materials included radioactive particles, and Parker's reaction indicated his strong disapproval and may have indicated that he had not known of the radioactive release before it occurred.

Frank Badgley and I were involved also in consulting work that Phil was doing for the Aluminum Company of America that was designing a production facility near Wenatchee, Washington. The problem was to determine how tall a stack had to be to reduce toxic effects of effluents to acceptable levels. Our work required measuring wind velocities and temperatures on the slopes bordering the Columbia River valley in order to estimate vertical profiles at the stack site. We encountered dramatic instances of chinook (foehn) winds on the mountain slopes west of the valley, indicating sharp temperature inversions at heights of 100–400 m above the factory site. Later, Frank and I also worked on predicting the temperature increase in the Nechako River in British Columbia that would result from construction of a dam.

In about 1951 the Weather Bureau initiated a field study under Ferguson Hall's direction to test claims advanced by cloud-seeding enthusiasts, specifically, that cloud seeding by aircraft in storms approaching the Pacific coast would result in increased snowfall in the Cascades. The project was based at Sand Point Naval Air Station and ran for two years or more; the primary result as I recall was that the banded structure of cloud systems approaching the coast introduced enough statistical uncertainty to mask any effects of cloud seeding on precipitation over the Cascades. This result strengthened the Weather Bureau's skeptical view of claims being made by cloud-seeding enthusiasts at the time.

Polarized views of cloud seeding had been demonstrated especially memorably at a meeting of the AMS that I attended in New York on January 25, 1950. Irving Langmuir, Nobel laureate in chemistry, presented a paper in which he claimed that periodic seeding in New Mexico had influenced precipitation over much of the United States. These claims were criticized by William Lewis of the Weather Bureau who had been assigned as an observer for the New Mexico study. Langmuir declined to discuss the issue, but following the session he and Ross Gunn, director of physical research for the Weather Bureau, became involved in a heated exchange that I and others who observed it thought was about to come to blows. Throughout the period of widespread enthusiasm for cloud seeding from the late 1940s to the 1970s the Weather Bureau maintained reservations based on scientific standards as well as on extensive experience in observing and analyzing weather; as a result it suffered much criticism, including some from influential members of Congress. Francis Reichelderfer, Harry Wexler, and Ross Gunn deserve special recognition for the Bureau's consistent stand.

In the fall of 1951 the department enrolled its first class of about 27 U.S. Air Force (AF) junior officers for meteorological training. To secure the needed space we moved from Thomson Hall to what was called the Anatomy Shack, the World War I temporary building now called Johnson Annex A. The group included Richard Johnston, Thomas Potter, Eugene Rasmussen, and Ernest Smerdon, who later earned Ph.D. degrees and went on to distinguished careers, and several others who earned M.S. degrees. In 1992 this class held its 40th reunion at Port Ludlow and invited Frank Badgley and me and our wives as guests. In each of the five or six years following 1951 similar groups were assigned to the UW for meteorological training. I do not remember these groups as clearly; but they included John Perry and Paul Try, who later returned to the UW for Ph.D. degrees [Perry (1966, R), Try (1972, L)]. After retiring from the AF Weather Service, Perry has served for many years as executive secretary of National Research Council committees and boards, and Try later held leadership positions at the World Meteorological Organization in Geneva, Switzerland, and in 1995 served as president of the AMS.

Also in 1951 I began to develop a six-credit summer lecture and field course in oceanographic meteorology (later air–sea interaction) at the university's Friday Harbor Oceanographic Laboratory (now Friday Harbor Laboratories). This represented continuation of studies that

Phil Church had conducted prior to World War II. For the first several years my course was coordinated with a course in physical oceanography, taught by Maurice Rattray, so that students taking both courses had a full academic program. Richard Fleming, first chairman of the new Department of Oceanography, was director of the laboratory. He encouraged use of the laboratory by the schools of forestry and fisheries, as well as the departments of botany, zoology, and geography; but after the first year or so the summer program had consolidated to courses and research in marine biology, physical oceanography, and air–sea interaction; and marine biology more and more came to dominate the program. Oceanography was dropped after 1959 when Fleming was replaced as laboratory director by zoologist Robert Fernald. The air–sea interaction course continued at Friday Harbor until the mid-1960s, after which it was offered on the Seattle campus.

At Friday Harbor meteorology students had opportunities to test and use simple instruments for measuring temperature, humidity, and velocity profiles over the sea and to calculate radiative and turbulence fluxes at the sea surface. In addition to teaching the course, I was able to do some research on optical measurement of lapse rate. These measurements were used to validate calculations of radiational cooling and fog formation that provided an explanation of the distinction between fog forming in contact with a warm water surface (often called "steam fog") and fog forming above a cold surface (often called "radiation fog"; Fleagle 1953). In each case fog formation depends on divergence of radiative flux close to the sea surface, though turbulence in the warm surface case makes the physical characteristics of the resulting fog quite different in the two cases. I have wondered why this work has been generally overlooked in later textbooks; is the result not believed by the authors, or is it considered obvious? I taught the summer course each year from 1951 to 1958; later Frank Badgley taught it for several years, James Deardorff for one, and I returned for the 1962 summer. Later it was taught on the Seattle campus, once by Mike Miyake and for many years by Kristina Katsaros.

As the meteorology program expanded Phil Church wanted to add a climatologist to restore balance between the fields identified in the department's name. He secured an appointment in 1953 for Konrad Buettner, an internationally known German bioclimatologist who had been brought to the United States by the military and was working on medical space research for the Air Force. His later research focused on radiation instrument development and field observations in the Carbon River valley of Mount Rainier.

The research program expanded further with a grant in 1953 from the Munitalp Foundation to develop a mobile observing facility for studies of cloud and aerosol physics, motivated by possible application to cloud seeding and to diffusion and transport of pollutants, and with a contract with the Geophysics Research Directorate (GRD) of the Air Force for studies of synoptic-scale motions (the first for which I was principal investigator). The Munitalp grant supported some of Buettner's research and led to Donald Fuquay's M.S. degree (1954, Bu), and the GRD contract supported M.S. degrees earned by Edwin Danielsen (1954, F), Yutaka Izumi (1955, F), and James Deardorff (1956, F).

In 1954 approval was obtained for a national search for a synoptic meteorologist. At that time the usual first step was to request recommendations from the leading people in the field, the "old boys' network." That procedure led to a suggestion by James Austin of MIT that we consider Richard (Dick) Reed, a member of their research faculty who had gotten his doctorate in 1949. I visited Dick in Cambridge on a trip east and was especially impressed by the breadth of his research interests that embraced stratospheric ozone as well as tropospheric weather analysis and forecasting. We invited him to visit the UW, at that time an unusual step in the recruiting process. The visit went well, and Dick was offered and accepted an appointment as assistant professor, bringing the faculty to five tenured and tenure-track faculty members.

Over the years Dick Reed has brought the department recognition as a national and international center for training and research in synoptic meteorology. He played the central role in discovery and analysis of the quasi-biennial oscillation, as well as in investigations of many tropospheric phenomena. Dick has received many awards including election to the National Academy of Sciences (NAS) in 1978, award of the Rossby Medal in 1989, the highest honor bestowed by the AMS, and award of AMS Honorary Membership in 1999. He has served in many positions of responsibility for the NAS, AMS, and the University Corporation for Atmospheric Research (UCAR).

With Dick Reed's addition to the faculty a Ph.D. program was proposed and approved by the administration and the Board of Regents. Dean Staley and Richard Hubley became the first Ph.D. students. Staley was awarded our first Ph.D. in 1956 (F) and was appointed an assistant professor at the University of Wisconsin; later he moved to the

University of Arizona where he taught and did research until his retirement several years ago. Dean Staley died in 2000. Dick Hubley also completed his degree in 1956 (Bu) and went to the National Science Foundation as manager of the NSF program of Arctic research in the International Geophysical Year (IGY). Later, while involved in IGY field work in Alaska, he tragically lost his life.

The number of graduate degrees earned increased markedly in the late 1950s. Dick Reed supervised M.S. degrees earned by William Tank in 1955, Thomas Potter in 1957, William Campbell and Richard Wilson in 1958, and Bruce Kunkel, Joseph O'Neal, and Nolan Williams in 1959. Frank Badgley supervised the M.S. degrees earned by Robert Baughman in 1958, and by Neal Barr, Chandran Kaimal, and Donald Stevens in 1959. Phil Church supervised the M.S. work of Archie Bloom in 1959. Ph.D. degrees were earned in 1958 by Robert Pyle (F), P.T. John (Ba), and Edwin Danielsen (R), and in 1959 by James Deardorff (F).

Departments somewhat similar to ours were being developed during the 1950s at The Pennsylvania State University, University of Wisconsin, The Florida State University, University of Arizona, Colorado State University, University of Michigan, Texas A&M University, and St. Louis University, while several other universities added more limited atmospheric components to existing departments. Faculty members from these universities and researchers at government laboratories met fairly regularly at meetings of the AMS and the American Geophysical Union (AGU) and through service together on committees, and their work was published for the most part in a small number of journals. Many of these scientists, representing a broad range of geophysical specialties, came to know each other both professionally and personally. Members of the community that evolved in this way gained broad views of the atmospheric sciences and had opportunities to influence a range of programs and activities. Today, similar communities of young atmospheric scientists tend to be more specialized, and their ranges of influence are probably more limited.

Planning for the IGY by the U.S. National Committee included Arctic research based on observations made on a station floating on the Arctic ice. At Phil Church's initiative the UW became the contractor for Station Alpha, operated during the IGY (July 1957 to December 1958). Norbert Untersteiner, an Austrian glaciologist, was appointed as chief scientist for the project, and a team of observers and operations people was recruited.[16] The focus of effort was on securing the observations needed to determine the energy budget for the Arctic region. Frank Badgely served as chief scientist on the station from September to December of 1957. Because Untersteiner had entered the country under a visitor visa, at the end of the IGY he was required to return to Austria.[17]

The Station Alpha project led to a continuing series of UW observational projects and to continuing research programs in glaciology and polar geophysics. "Project Husky," begun in 1958, supported graduate student research. In April 1960, as part of Husky, I participated in "The Lead Experiment," an experiment designed to estimate thermal losses occurring in open leads in the Arctic ice. The measurements and the method of calculation were those used in the summer course in air–sea interaction. A pond 25 m or so in diameter was formed on the ice by pumping water from below the ice. Vertical profiles of temperature, humidity, and wind velocity over the pond were used to calculate evaporation and heat transfer as a basis for estimating losses from open leads distributed over the Arctic Ocean. The observations provided the data for Mike Miyake's M.S. thesis in 1961 (Ba). The IGY also provided support for establishing a research station on the Blue Glacier on Mt. Olympus and for continuing studies there led by Ed LaChappelle. In later years many other UW glaciological and Arctic studies have been carried out within atmospheric sciences, the Polar Science Center, and the geophysics program; and the UW has played a leading role in international polar programs. The IGY also stimulated broader faculty interest in geophysics and led to Phil Church's initiative in forming a faculty committee that met to discuss interdisciplinary research projects.

In the summer of 1957 I attended the 12th General Assembly of the International Union of Geodesy and Geophysics (IUGG) in Toronto, Ontario, Canada, where I met Joost Businger who had spent the previous

---

[16] Untersteiner had been awarded a postdoctoral fellowship by Robert Sharp at the California Institute of Technology to pursue glaciological studies as part of the IGY, but the fellowship had to be withdrawn when funds were not available. Sharp called this situation to the attention of IGY headquarters at NSF where Dick Hubley was working, and Untersteiner was proposed as chief scientist for Station Alpha and appointed by Church. He arrived at the UW in February 1957 and set to work immediately to establish the station.

[17] See chapter 8 for an account of Untersteiner's later role in the department.

year working with Verner Suomi and with Dean Staley at the University of Wisconsin. On a bus trip to Niagara Falls we discussed topics ranging from atmospheric physics to science organizations and the future of the atmospheric sciences. I was especially impressed that Joost had developed a mathematical model of the boundary layer that could be integrated to describe the growth of the layer as a function of observable quantities. Our meeting in Toronto led, a year later, to Joost's joining the department as an assistant professor, further strengthening our research and teaching in boundary layer processes. Businger's first Ph.D. student was Chandran Kaimal, who earned his degree in 1961.

Joost Businger has been a leader in boundary layer research; his work has included a central role in development of the sonic anemometer, leadership in the 1967–68 Kansas field experiments, and papers clarifying characteristics of the stable and unstable boundary layers. He served as chair of the University Senate in 1976–77 and of the department in 1982–83. His awards include the 1978 AMS Second Half Century Award, election in 1980 as a Fellow of the Royal Dutch Academy of Science, and election in 2001 to the U.S. National Academy of Engineering.

CHAPTER 3

Political Events at the University of Washington

Beginning with the Canwell hearings of 1948, restraints on academic freedom at the UW extended into the 1950s, exposed contrasting views of the university's purpose and responsibilities, damaged its reputation, disrupted normal campus processes, and in my judgment jeopardized the future of the institution.[18] Fortunately, later events helped to arrest this decline and contributed to a period of development ultimately leading the UW to positions of leadership in many academic fields. The periods of decline and development have been part of corresponding events played on the larger national and international stages; they have been important drivers in the evolution of the atmospheric sciences over the past half century.

When I arrived at the UW in 1948, as noted in chapter 1, the faculty Tenure Committee was investigating charges brought by the university administration against faculty members accused of being Communists, and tensions were high. Phil Church introduced me to a member of the Tenure Committee, Thomas G. (Tommy) Thompson, chemical oceanographer, progenitor of the Oceanography Department, and the only UW faculty member who had been elected to the National Academy of Sciences. Along with the majority of the committee, he voted against the administration's position. However, early in 1949 President Raymond Allen, citing critical comments of the Tenure Committee as justification, recommended firing three faculty members and putting three others on probation. The Board of Regents carried out these recommendations, effectively ending the careers of the directly affected faculty members and chilling freedom of expression for the campus as a whole. These actions were protested by many prominent liberals around the country. In this time of great tension a petition protesting the actions of Allen and the Regents was circulated among the faculty and signed by 103, including me. Given the attitude of the

---

[18] These events are described by Sanders (1979).

administration, which many senior faculty including Phil Church supported, many of the signers had to recognize that the petition might well lead to an end to their employment at the university. I felt that if that was actually the case, I would not want to remain.

In another case the Canwell Committee accused Philosophy Professor Melvin Rader of being a Communist and of having attended a 6-week Party workshop in New York, both of which he denied. Independent investigation revealed that the charges were false and were based on altered and mysteriously missing records and on testimony of a paid witness. The Canwell Committee and the witness were discredited, but a year and a half of silence passed before the charges were finally publicly dismissed by President Allen. Edwin Guthman, a *Seattle Times* reporter, was awarded the Pulitzer Prize for best national reporting in 1949 for his role in exposing the criminal deception.[19] The history of this case conveys a sense of the bitter taste of polarized campus life at the time.

National politics during the first half of the 1950s were dominated by Senator McCarthy's charges of communism in government and investigations by Senate and House committees of university faculty and others suspected of liberal views or associations. A movie, *Communism on the Map*, produced with support by right wing individuals and corporations including The Boeing Company, frightened audiences with visions of an implacable evil force covering the earth. A personal friend of mine, Abraham Keller, a romance languages faculty member who had been a member of the Communist Party before and during the Second World War, was unwilling to name party members when called before the House Committee on Un-American Activities. He narrowly escaped being charged with contempt of Congress and losing his faculty position. In 1955 an invitation to J. Robert Oppenheimer to give a Walker–Ames (endowed) lecture, was withdrawn by President Henry Schmitz. When Giovanni Costigan, widely known and respected as a liberal UW professor of history, criticized the accuracy of *Communism on the Map*, he was viciously attacked in the media. I wrote a letter published in one of the Seattle papers, or perhaps the *University Daily*, defending Costigan and pointing out that he was serving his professional responsibility by correcting historical errors. I also spoke before several small groups criticizing actions that I saw as threatening free speech.

---

[19] For a detailed account see Rader (1969).

At these meetings I encountered emotional opposition but nothing more. In later years when I encountered Costigan, he often expressed his appreciation for my letter defending him. This surprised me because I had not known Costigan before and had assumed that many others also must have come to his defense.[20]

Reactions to the Oppenheimer ban included cancellations of visits by several prominent individuals. A letter to President Schmitz signed by seven eminent biochemists stated that "it seems to us that you have clearly placed the University of Washington outside the community of scholars."[21] A major conference on molecular biology to be held on the campus collapsed. On the other hand, the ban was strongly supported by the two Seattle papers and by influential persons on and off the campus. The central issue was clearly posed. Should the role of the university be to carry out functions prescribed and limited by the president, or should the university operate as a community of scholars? In the answer to this question lay starkly contrasting futures for the university and the state.

The UW chapter of the American Association of University Professors, with a membership of more than 300, disapproved Schmitz's decision by a ratio of 3 to 1. The faculty Senate, after long debate, adopted a resolution stating that it "regrets the damage done to the university by the decision in the Oppenheimer case." President Schmitz tried to repair the damage, while maintaining that his decision had been in the best interest of the university. He approved Oppenheimer's appearance on the campus the next year at an international meeting on theoretical physics, met with Oppenheimer, and is reported to have attended Oppenheimer's talk. But the stain remained, and was visible to those who valued academic freedom.

A loyalty oath had been among papers perfunctorily signed by all state employees at the beginning of their employment. In 1955 a new,

---

[20] Many years later Marianne and I enrolled in Costigan's final courses on English history and the British university, given in the summer of 1975 at the Universities of London, Oxford, and Cambridge, where we lived in student rooms and used the college libraries and refectories. We especially admired and cherished Costigan's uncompromising scholarship, his determined advocacy of humanitarian goals, and his gentle manner.

[21] *Seattle Times*, March 23, 1955 quoting a letter of March 14, 1955 to President Henry Schmitz signed by seven biochemists including at least two Nobel Laureates.

more specific oath was introduced with little fanfare; I along with nearly all the faculty signed without much thought. However, Max Savelle and Howard Nostrand, senior members of the history and romance languages faculties, respectively, obtained an injunction preventing enforcement of the law requiring the oath. A legal case challenging the required oath as unconstitutional was instituted by a group of faculty, graduate students, and staff. Joost Businger, Arnold Hansen (research meteorologist), and I were among the 63 plaintiffs. Each of us wrote a letter to the university president withdrawing the oath that we had signed, and we wrote a statement of why we had taken this action. I emphasized that prohibition of association with Communists would exclude broad scientific communication and participation in international scientific meetings, which were part of my professional responsibilities.

The case worked its way through the courts during the late 1950s and early 1960s and was argued before the U.S. Supreme Court in the fall of 1963 by Arval Morris of the UW Law School and Kenneth MacDonald, a liberal Seattle lawyer.[22] Because I was spending the 1963–64 year in Washington, D.C., I was able to attend the hearing, the only one of the plaintiffs in attendance. It was an inspiring experience that enhanced my already great respect for Morris and MacDonald. In the spring of 1964, about eight years after the legal case had been initiated, the Court reversed the prior decision of the U.S. District Court and declared the oath unconstitutional by a 7 to 2 vote. Justice Byron White wrote the majority opinion. The decision was based in part on the need for faculty to communicate freely with scientists and scholars in Communist countries. This decision invalidated similar oath requirements of certain other states. I felt that the Supreme Court decision contributed substantially toward the UW becoming a major national university and strengthened civil liberties nationally.

By 1964 other currents also had begun to move the university in a positive direction. When Charles Odegaard became president in 1958 he appointed respected faculty to administrative positions and acted in other ways to raise standards, increase enrollment, raise faculty salaries, and recruit and retain outstanding faculty. There was a new tone on the campus. President Odegaard was able to gain state support for regular budget increases and to insulate the campus from legisla-

---

[22] The many steps leading to this final hearing are described in Sanders (1979).

tive meddling. Also, Odegaard's early presidency coincided with a period of increasing federal support of universities stemming from U.S. response to the launch of *Sputnik*; and the UW, with substantial aid from Senators Magnuson and Jackson, was highly successful in securing grants for research and for new facilities. During the 1960s these factors together enabled many UW departments, especially in the sciences and medicine, to gain recognition as national leaders in their fields.

This generative era ended rather abruptly in the late 1960s as a result of the Vietnam War. Federal university support was curtailed, and the university was beleaguered by student activism and protests. Bombings and other violent actions occurred on campuses across the nation, and at the UW the Administration Building was severely damaged by a bomb. In reaction to the killing of students at Kent State University by the National Guard, UW students and faculty forced closure of the university, and preparations were started to organize a delegation to go to Washington in protest. As chair of the department I called a meeting of faculty and students to consider departmental action; after animated discussion we voted to send Joost Businger to Washington to represent the department. The UW delegation consisted of Joost, Larry Wilets of the Physics Department, and four students who had collected thousands of signatures that were hand-delivered to President Nixon. On the trip back from Washington Joost sat with Senator Jackson and discussed with him some of the views being expressed on the campus. On the academic front standards fell, as did the morale of faculty and students. The university faced a stressful period that extended to the end of the Nixon administration.

These stresses were expressed to me in a quieter and more personal way. In the mid-1960s I had been appointed to the Subcommittee on Air–Sea Interaction of the NATO Science Committee; this led to recruiting a few students on NATO scholarships for our summer course in air–sea interaction.[23] After several years as a member of the subcommittee I became aware that my security clearance was being reviewed. Letters I had written protesting the Vietnam War had been published; and I had signed many petitions of protest, so investigators had lots of new fodder to feed into the security mill. It was a time of

---

[23] Later, under Kristina Katsaros's leadership these contacts broadened and were quite productive.

growing evidence of subversion of the FBI and the CIA and of reports of a White House "enemies list." I became aware that, though I had been proposed for membership on the National Committee on Oceans and Atmosphere (NACOA), my appointment had not been approved. I decided that, while there might be a certain distinction to inclusion on the enemies list, I did not want to jeopardize the embryonic UW program of scholarships for NATO students. In any case I was thoroughly sickened by actions of the Nixon administration, so resigned from the NATO subcommittee.

CHAPTER 4

Windows on National and International Affairs

When I was elected a member of the Council of the AMS in 1957, I was exposed more directly than before to national meteorological affairs. From that time forward, I can now recognize, my interests gradually broadened to include institutional matters and science policy, and this broadening is reflected clearly in the chronological list of my publications.[24] However, the central motivation throughout my career has been to strengthen the department and to move the UW toward becoming a leading center of education and research in the atmospheric sciences. I have considered activities relating to policy as important contributors in that quest.

During my term on the council it was concerned with the highly charged issue of dealing with Irving Krick's extravagant claims of success in long-range forecasting that many felt violated the Society's Code of Ethics. Krick was the colorful former head of the Meteorology Department at the California Institute of Technology who had many influential friends and associates including Robert Millikan (former president of Cal Tech) and Theodore von Karman (world famous hydrodynamicist and aeronautical engineer). Although some councilors wanted to publicly expel Krick from the Society, others recognized that this step could be costly and that it might not be possible legally if Krick chose to challenge the action. I recall the tension of the May 6, 1958 council meeting, chaired by Henry Harrison in Washington, D.C., at which Krick offered his resignation and the relief with which it was accepted. Minutes of the meeting refer somewhat cryptically and cautiously to resignation of a member "who had been charged with a possible breach of the Code of Ethics." Officers who led the council through this difficult issue were President Sverre Petterssen, Vice President Henry Harrison, and Secretary Tom Malone. Malone especially deserves credit for behind the scene negotiations that secured the

---

[24] Bibliography for Robert G. Fleagle

resignation. The council also was involved at that time in acquiring the historic house on Beacon Hill in Boston that became the Society's headquarters, acquired through the primary efforts of Executive Secretary Kenneth Spengler with help from Frances Ashley and David Landrigan.

Early in 1958 the report of the Committee on Meteorology of the National Academy of Sciences recommended creation of a national institute of atmospheric research, to be supported by the National Science Foundation, and set in motion major changes that have affected the field in fundamental ways ever since.[25] The committee's report was welcomed promptly by key officials, and a "Blue Book" outlining the functions and structure of the national institute was prepared through Tom Malone's special efforts. The initial institutional step was formation of the University Committee on Atmospheric Research by the representatives of 14 universities at a meeting held at UCLA in February 1958. At Phil Church's request, I represented the UW at that meeting. Although parochial interests were very visible and vocal at the meeting, the wise chairmanship of MIT's Henry Houghton helped to project a vision of exciting science and expanding value to society, dependent on cooperation among the universities. Competition between the universities and the national institute emerged immediately as an issue and was resolved by agreement that staff of the institute would not be recruited from tenured faculty. That issue was to reemerge from time to time in many guises. Over the following year the committee worked out a variety of issues, procedural and substantive; and in March 1959 the committee became the University Corporation for Atmospheric Research (UCAR) and set about creating the National Center for Atmospheric Research (NCAR).

I spent the year 1958–59 at the Imperial College in London, on sabbatical leave from the UW and as a National Science Foundation Senior Postdoctoral Fellow. The Department of Meteorology at Imperial College was then the chief meteorological research center in Britain, and interests of its faculty covered a broad range: Eric Eady in dynamic me-

---

[25] The committee was appointed at the request of the Chief of the Weather Bureau Francis W. Reichelderfer. Cochairmen of the committee were Carl-Gustav Rossby and Lloyd V. Berkner and members were Henry Booker, Horace R. Byers, Jule G. Charney, Carl Eckart, Paul Klopsteg, Thomas F. Malone, and Edward Teller.

teorology, B.J. (later Sir John) Mason and Frank Ludlam in cloud physics, Department Chairman Peter Sheppard in boundary layer turbulence, Richard Goody in radiation, Richard Scorer in mesoscale dynamics. Henry Charnock was in residence for at least part of the year and I believe gave a course in oceanography. The year proved interesting and worthwhile, even though the department was disbanding at that time. Goody left for Harvard as I arrived (I inherited his new, elegant but isolated office on the top floor), Eady came to the college for his class lectures but otherwise was not present, Ludlam, who had moved to Silwood Park Farm, came in only about once a week, and Scorer seemed aloof or alienated. Nevertheless, Sheppard, Mason, Charnock, John Green, Peter Saunders, and Ludlam were doing interesting work and made morning and afternoon tea lively times. Sheppard and Mason had sharply different views about the department and other matters. Their differences were expressed with equal parts of British acerbity and civility. Many years later, shortly before his death, Sheppard said to me that, despite Mason's "sharp elbows," he had been especially thoughtful and kind in a difficult time.

A casual event made me recognize a profound difference between citizens of the United States and the USSR. On December 18, 1958 the United States succeeded in launching its third earth satellite, the first to send voice messages to earth and the heaviest payload placed in orbit to that date. The following morning I was working at my desk when a Soviet oceanographer who was a visitor at Imperial College appeared in the doorway and announced in a very formal voice, "I congratulate you on your satellite success." I immediately responded, "I had nothing to do with it, don't congratulate me." Then I realized that, whereas I was simply disclaiming credit I didn't deserve as an individual, he identified himself as part of all his nation's achievements and was crediting me with a similar reaction. I regretted not having been more sensitive and appreciative of his intended courtesy.

Sheppard arranged for me to give lectures at Manchester and Cambridge on my work on baroclinic instability (Fleagle 1957) that had received honorable mention for the first Sir Napier Shaw Prize [awarded to Norman Phillips for his seminal paper on a numerical model of the general circulation (Phillips 1956)], and in many other ways he made the year stimulating and enjoyable. At Cambridge Sir G.I. Taylor attended my lecture and asked how I had found a frequency equation without introducing boundary conditions. I explained that this had been possible by defining the z-direction of a Cartesian coordinate sys-

tem as normal to the streamsurface of maximum slope in the midtro-
posphere, whose existence and slope were revealed by my earlier studies of midlatitude storms. Upon applying the equations of motion on
this streamsurface the normal velocity and the divergence of normal
velocity both vanish. This yields a frequency equation that includes the
streamsurface slope as a parameter that can be eliminated using mean
temperature gradients in the vertical and north–south directions. I
don't know whether this answer fully satisfied Sir Geoffrey, but he did
not pursue the question further.

I was invited to attend monthly meetings of the Meteorology Club,
an organization of distinguished scientists and professionals whose
regular meetings continue to the present. These meetings provided a
close view of British manners and traditions and introduced me to
Reginald Sutcliffe, J.M. Stagg, and others whom I otherwise would
have known only through their publications. On one occasion I met
P.M.S. Blackett (Nobel laureate in physics) at lunch in the Senior
Common Room and was immediately impressed by his broad interests
and warm humanity. Meetings of the Royal Meteorological Society
were always stimulating for the provocative, animated question periods, in distinct contrast with many meetings of the AMS. On a visit to
the Institute of Meteorology in Stockholm I met Bert Bolin, Pierre We-
lander, Claes Rooth, and George Witt, and renewed acquaintance with
Phil Thompson, who was also a visitor. Bert Bolin had succeeded Carl
Rossby as director of the institute at Rossby's death, and the institute
had become an exciting center of research and an international leader
in the atmospheric sciences.

By the time I had returned from sabbatical, growth of interest in
atmospheric processes had been reflected in adoption of the term "atmospheric sciences" as describing the field, especially by the National
Academy of Sciences (NAS) and the National Science Foundation
(NSF). I was appointed to the Panel on Air–Sea Interaction, a new joint
panel of the NAS Committees on Atmospheric Sciences and Oceanography (referred to, respectively, as NAS/CAS and NASCO). Other members of the panel were George Benton (chairman), Raymond Montgomery, Dale Leipper, Herbert Riehl, Norris Rakestraw, William S.
Richardson, and James Snodgrass. Initially, Joanne Malkus (later
Simpson) was appointed to the panel but withdrew, probably due to the
press of family responsibilities. At about the same time I was appointed
to the Advisory Committee of the Woods Hole Oceanographic Institution. The work of the Benton panel began for me an association with the

National Academy that spanned more than two decades. I remember especially the panel meeting at College Station, Texas, on January 20, 1961, at which we listened with excitement and hope to President Kennedy's inaugural speech. The panel's recommendation of field experiments to compare fluxes of heat, water vapor, and momentum measured by independent methods led to my appointment as chairman of an Academy panel charged with planning part of the International Indian Ocean Expedition. Later, the Barbados Oceanographic and Meteorological Experiment (1969) was carried out in response to recommendations of the Benton panel. I served as principal investigator of UW activities in each of these international programs.

President Kennedy came into office committed to encouraging stronger support for science. He relied on his Science Advisor, Jerome Wiesner, in selecting key officers for science agencies; science assumed an enhanced profile; and important science and technology initiatives were introduced. At Wiesner's request a small group was convened by Bruno Rossi (MIT physicist) to consider scientific initiatives that would foster closer cooperation among nations and help to reduce tensions between the United States and the USSR. The group included Thomas F. Malone, Jule Charney, and Richard Goody.[26] Following this meeting an ad hoc panel of the President's Science Advisory Committee (PSAC) was convened to examine possible atmospheric initiatives; it was attended by Wiesner, Detlev Bronk (NAS president), Arthur Schlesinger, Jr. (special assistant to the president who took copious notes), other senior officials, and Tom Malone, who described to the panel possible initiatives in the atmospheric sciences.[27] The report of this panel urged research to improve weather forecasting and included the possibility that weather control might become feasible in the future. In response to that report Wiesner requested the National Academy to undertake a study of research needs and opportunities in the atmospheric sciences. The Academy President, Detlev Bronk, asked Tom Malone to chair the study, but he deferred to Sverre Petterssen, chairman of the Geophysical Sciences Department at the University of Chicago, who was appointed as study

---

[26] Malone was director of Research at Traveller's Insurance Co., chairman of NAS/CAS, and former associate professor of meteorology at MIT. Charney was professor of meteorology at MIT and Goody Professor of Meteorology at Harvard.

[27] These two meetings were reported to me much later by Tom Malone.

director. I was asked to chair a three-day conference on Atmospheric Structures and Circulation, one of six conferences that contributed to the study. By September 1961 the study had been completed and a report prepared. The report recommended trebling of research investment over the next decade and creation of "centers of excellence in which aeronomy is merged with meteorology into a unified atmospheric science program which in turn becomes an integral part of graduate programs in geophysics, closely associated with first-rate work in classical physics" (NAS/CAS 1962). This somewhat prolix phrasing reflected the ferment stirring within the atmospheric sciences community and set a challenging goal for future research and graduate education.

The report also recommended development of a truly global observing system and an international science program. These recommendations, together with the report of the ad hoc PSAC panel, became the basis for President Kennedy's proposal to the 16th General Assembly of the United Nations on September 25, 1961 for "cooperative efforts between all nations in weather prediction and eventually in weather control" (Kennedy 1962). Inclusion of the phrase "eventually in weather control" in his speech and in the resulting U.N. resolution reflected widespread belief in the virtually unlimited power of technology, belief fostered by wartime technological successes, especially development of the atomic bomb.[28] No questions were raised publicly before or after the speech as to whether weather control was desirable—if it could be achieved.

Kennedy's proposal was endorsed by the U.N. on December 20, 1961, setting in motion a long series of national and international actions. These events were linked to major issues facing the Kennedy administration. In response to successes of the Soviet Union in space, Kennedy, in a speech before Congress on May 25, 1961, had launched the decade-long effort to land a man on the moon. The President proposed the international weather program, at least in part, to encourage international cooperation as a counter to U.S.–Soviet competition in nuclear weapons and in space. And Jerome Wiesner and PSAC recognized weather satellites as providing a desirable alternative to the vague goals and technological uncertainties of manned exploration of space.

---

[28] Malone assumes that it was Schlesinger who was responsible for inclusion of the words "eventually in weather control" in Kennedy's U.N. speech. It probably was at least reviewed by Wiesner.

Soon after completion of the study I became a member of NAS/CAS, so was involved in responding to the report, as well as in preparing it, a practice more acceptable then than it would be now. The committee endorsed the report, and it was transmitted through Academy President Bronk to the White House. The endorsement referred to the fact that the report directed less attention to aeronomy than to meteorology. When Wiesner requested a more complete study of aeronomy, committee member Gordon Little led an additional study that was completed early in 1962. Although Petterssen wanted his report to stand alone, the committee felt it essential that the Little report be included. The original report became Parts I and II and the Little report became Part III of the published report. The expeditious production and handling of this report stands in marked contrast to that of many National Academy reports in later years.

Soon after the study was completed Jule Charney proposed to NAS/CAS an international research program to test general circulation models. Charney viewed this as an essential step toward the international science program proposed by President Kennedy and endorsed by the United Nations. Charney's proposal was received with enthusiasm by NAS/CAS and the meteorological community.[29] This story is continued in chapter 7.

Also in 1961 Sverre Petterssen, over lunch at the Willard Hotel in Washington, offered me a University of Chicago professorship at a salary about 50% above my UW salary. Even though I preferred to stay at the UW, I felt that in view of the uncertain fate of the loyalty oath case I should consider the offer. On a visit to Chicago I enjoyed association with George Platzman and had a pleasant talk with President George Beadle. He had spent a sabbatical at Oxford the year before I had been in London, experiences that had left both of us committed Anglophiles. I was interested in the concept of a unified department of geophysical sciences; but I was not convinced that a forced marriage of the geology and meteorology departments, exemplified at Chicago, was the way to accomplish it.

I reported on my Chicago visit to Dean Solomon Katz and empha-

---

[29] The NAS/CAS report and Charney's proposal can be seen as built on the successes of the Thunderstorm Project (1946–48), the largest meteorological research project to that time (Byers and Braham 1949) and the much larger International Geophysical Year (1957–58).

sized that the UW had an excellent opportunity to strengthen the atmospheric and related sciences and to become a national leader in the geophysical sciences. If we could respond to the opportunity I would be glad to stay at the UW. I would take my chances on the loyalty oath. Discussion led to his agreeing to broadening the scope of the department by new appointments in cloud physics, dynamics, and aeronomy, changing the name to Atmospheric Sciences, and appointing an interdepartmental committee to propose further steps in development of the geophysical sciences. Katz agreed to my suggestion that Joost Businger should be asked to serve as chairman of the committee. He said that he could not match the Chicago salary offer but could offer an increase of about 30%. These seemed to me very encouraging responses, and I agreed to stay at the UW and left the dean's office with a heady sense of optimism. Phil Church readily agreed to the various changes, though he may have been less than enthusiastic about the role of the geophysics committee.

The ad hoc Committee for Geophysical Sciences, consisting of Joost Businger and me (atmospheric sciences), Maurice Rattray and Joe Creager (oceanography), Peter Misch (geology), Kenneth Clark (physics), Theodor Jacobsen (astronomy), Myron Swarm (electrical engineering), George Halsey (chemistry; later replaced by Arthur Fairhall), and William Phillips (associate dean), was appointed and set to work immediately. Our discussions ranged widely; we all wanted to encourage and strengthen geophysics at the UW; but individual views differed on definitions and the institutional structure we would propose and on many other aspects. Fortunately, Phillips as secretary was able to produce minutes of the meetings that revealed more coherence than was evident to most of us at the time of the meeting.

In the spring of 1963 we proposed a graduate program that encountered both interest and resistance within the general faculty. Opposition was centered especially within the geology department.

A committee of the NAS concerned with new initiatives in geophysics education, chaired by Richard Goody of Harvard, expressed interest in our proposal; but two years or more passed before the Graduate School was able to approve the new program and graduate students were admitted to the program. The delay may have been a consequence, at least in part, of my spending the year 1963–64 in Washington, D.C. and Joost's spending the year 1965–66 in Australia. Approval was finally gained on a number of new courses and on appointment of a Geophysics faculty consisting of joint appointments with existing related departments and new appointments in areas of geophysics not represented in the existing departments. In 1967 Ken Clark, specialist in auroral physics, became the first chairman of the Geophysics Program, Businger and Untersteiner each received joint appointments in Atmospheric Sciences and Geophysics, and other joint appointments were made. I became chairman of Atmospheric Sciences at that time; otherwise, I probably also would have joined Geophysics half time.

I supervised the first Geophysics Ph.D. student, Robert A. Brown, who earned his degree in 1969. After a postdoctoral year at NCAR he returned to the UW where he has done extensive research on the theory of helical boundary layer circulations and remote sensing of the ocean surface and has collaborated in a number of international projects. Here at the UW, Brown has taught graduate courses in boundary layer and geophysical fluid dynamics, published two textbooks, and supervised the theses of a series of students. The first Ph.D. awarded under his direction was earned by Gad Levy in 1987.

During the early 1960s much of my attention, and of Joost Businger's, was directed toward the writing and publication of *An Introduction to Atmospheric Physics*. We both felt there was need for a text that would encourage the interest of physics students in the atmospheric sciences and would provide for students of the atmospheric sciences the fundamentals of general physics. Some years earlier I had met and corresponded with Kurt Jacoby of Academic Press, who expressed interest in a book along those lines. I believe he was a refugee from Hitler's Germany who had owned Akademische Verlag before the war, and had come to New York where he founded Academic Press. The task of writing and rewriting the manuscript was engrossing, and I believe it substantially improved my teaching. Academic Press provided help in drafting of figures, paid for a typist, and tried to meet our requests in other ways, personal attention that would be unusual today. The book was published in September 1963 (Fleagle and Businger 1963). After several reprintings, a second edition was published in 1980 (Fleagle and Businger 1980); over the years sales, never large, have remained at a fairly stable rate. We take this to indicate that the book still provides a sound and attractive introduction to atmospheric physics.

CHAPTER 5

The Other Side of the Window

One morning early in December 1962 I received a call from David Beckler, executive assistant to Jerome Wiesner, who asked if I might be interested in joining Wiesner's staff in the Office of Science and Technology (OST), part of the Executive Office of the President. The next week I traveled to Washington for discussions with Wiesner, Beckler, Edward Wenk, and David Z. Robinson. Wenk was staff specialist for oceanography as well as executive secretary of the Federal Council for Science and Technology, and Robinson was staff specialist for the physical sciences. They and about 14 other specialists were assistants to Wiesner, the president's science advisor. The position proposed for me was as staff specialist in the atmospheric sciences, a new position created in response to the emphasis being placed on the atmospheric sciences and the growth of atmospheric research budgets of the various agencies. Responsibilities would include becoming informed on research and development programs of the agencies, attending meetings of the President's Science Advisory Committee (PSAC) and serving as staff secretary for the PSAC Panel on the Atmospheric Sciences (chaired by John Tukey and known as the "Tukey Panel"). I also would serve as OST observer for the Interdepartmental Committee for Atmospheric Sciences (ICAS), the committee of the Federal Council charged with coordinating atmospheric research supported by the various agencies. Perhaps most important, I would act as liaison or consultant to the Bureau of the Budget on atmospheric issues. I also would attend meetings of the Interagency Committee for Oceanography (ICO) and assist Ed Wenk in advising on ocean research programs and budgets and preparing ICO reports.

I was intrigued by what I had heard, but replied that I was committed to teaching through the academic year and to supervising David Houghton's thesis as it approached completion. Wiesner agreed to my deferring the move to Washington until June and to my continuing to work with Dave Houghton if he could come to Washington. He also agreed on my completing my commitments for the IUGG meeting to be held in Berkeley in August, but said that I should resign from the Woods Hole Advisory Committee to avoid conflict of interest. I brought up my participation as a plaintiff in the UW loyalty oath case that might come before the Supreme Court during the next year, and asked if that presented a problem. Dave Beckler asked if I was a Communist, to which I simply said "no." Ed Wenk said that as a citizen I had every right to be a plaintiff; and Jerry Wiesner simply passed on to other matters.[30] I felt that OST had passed an important test.

Two weeks later I received a formal letter of appointment from Wiesner, and over the following six months I made regular two-day trips to Washington to serve as a consultant and to learn more about OST procedures, and in June moved with my family to Washington. My security clearance had been approved by the time I reached Washington, but according to the FBI file I obtained much later under the Freedom of Information Act it had been delayed by my record of liberal activities, and prodding by the White House staff may have been required.

I arrived at OST within a few days of President Kennedy's speech at American University in which he launched his proposal for achieving limitations on nuclear testing. This was followed by negotiations during the summer on tests in the atmosphere and in space, and the treaty banning these tests became effective on October 10.[31] I had no part in the negotiations, but I felt inspired by being close to these epochal events, and I was infused with increased enthusiasm for working for this president.

Dave Houghton moved to Washington for the summer, and we scheduled regular times for discussing his thesis. I recall one staff member saying to me that our afternoon discussions added an academic dimension to OST, which he thought provided a desirable ambiance.

However, I soon learned that life around the White House con-

---

[30] Wenk and I discovered that we had both been members of the undergraduate class of 1940 at Johns Hopkins, he in Civil Engineering, I in Physics, but had not known each other. We became good friends, especially after he joined the UW faculty in 1970 to head a new program in the Social Management of Technology.

[31] It was probably at this time that I first learned of Wiesner's earlier conversation with Kennedy one night while rain fell in the rose garden and ran down the windows of the oval office. Kennedy asked Jerry if fallout from bomb tests was in that rain, to which the reply of course was "yes." The American University speech followed with the memorable words, "For in the final analysis, our most basic common link is that we all inhabit this small planet. We all breathe the same air. We all cherish our children's future, and we are all mortal."

trasted markedly with academic life. Here it was necessary to respond quickly to unending streams of requests and challenges. I came to appreciate how thin was the basis for many decisions made within the agencies and in the White House. This placed an extraordinary premium on the background and judgment of key individuals, and I came to realize how important had been Wiesner's role in appointment of officers in the science agencies. In this respect the Kennedy administration set a standard that may not have been matched before or since. Attendance at PSAC meetings provided the opportunity to observe and listen to such luminaries as Isador Rabi, Richard Garwin, Harvey Brooks, John Bardeen, Paul Doty, John Tukey, Wolfgang Panofsky, Melvin Calvin, Donald Hornig, Victor Weisskopf, Sidney Drell, and others, as well as Jerry Wiesner, as they responded to issues ranging from design and location of the next particle accelerator, to the value of the Apollo Program, to improving high school science curricula, world hunger, and many more. PSAC approval was essential for undertaking new science or technology projects requiring substantial budgets. Wiesner, as PSAC chairman, was responsible for leading this group, for the outcome of its deliberations, and for communicating its conclusions to the president. The OST staff had learned that on PSAC days Wiesner was uncharacteristically tense and impatient.

Late in the summer Jerry asked me to visit the Travellers Research Center in Hartford, Connecticut, and its President, Robert M. White. I had known Bob White for more than 15 years, though not well; and I had high regard for his scientific and administrative achievements. The visit strengthened this impression, and I reported positively on the job White was doing. He was being considered for appointment as chief of the Weather Bureau, replacing Francis Reichelderfer, who was retiring. I was glad to support the appointment. When Bob White was appointed and took office early in the fall I felt assured that the Weather Bureau was in good hands and that it would follow a vigorous and scientifically sound course.

I attended a meeting of the Interagency Committee for Oceanography (ICO) at Woods Hole in the last week of August. As a result I missed the 1963 Civil Rights March and Martin Luther King's speech at the Lincoln Memorial. The Woods Hole meeting was especially memorable because during the meeting Jerry Wiesner was involved in a boating accident off Martha's Vineyard in which he nearly lost his life. News of that event disrupted the meeting, and I have no record or recollection of the agenda or of any decisions reached there.

Back in Washington I learned that, though there was general approval within OST and the Bureau of the Budget (BoB) for increased NSF support for the atmospheric sciences, NCAR was being criticized from several directions. The PSAC Tukey Panel came away from a visit to NCAR less than enthusiastic about the quality of the growing staff, and it criticized the emphasis that Director Walter Roberts had placed on individual research and consequent failure to focus on major research goals. University scientists criticized the rapid growth of NCAR budgets (the chronic issue of competition between NCAR and the universities); and J. Herbert Hollomon, Assistant Secretary of Commerce for Science and Technology, stated at every opportunity that NCAR was wasting resources and should be transferred from NSF to Commerce. Hollomon was a bright, energetic, and influential figure who had Wiesner's confidence; and his statements disturbed NSF officials and the UCAR–NCAR community. However, Hollomon's efforts to take over NCAR received no support from Wiesner or PSAC. NCAR's response to criticism of the Tukey Panel was slow and did not fully satisfy the panel.[32] With respect to the rapid growth of NCAR budgets, my responsibility was to describe to PSAC, the Federal Council, and the BoB [predecessor of the Office of Management and Budget (OMB)] NCAR's central functions, especially its assistance to university research, and to identify good and important research being done there while recognizing the criticisms it was receiving.

During the fall negotiations between the Department of Commerce and NASA led to an agreement specifying the two agencies' responsibilities for operational and research weather satellites. Hollomon pushed this agreement vigorously, and I was glad to support it. Hollomon also proposed establishment of the Office of the Federal Coordinator for Meteorology to coordinate the meteorological services of the Weather Bureau, Air Force, Army, Navy, and FAA. Although I supported the objective, I was concerned that the name Federal Coordinator for Meteorology could imply inclusion of the atmospheric research programs funded by NSF and other agencies. I pointed this out at a

---

[32] After returning to the UW in 1964 I became involved in internal efforts to strengthen the UCAR–NCAR enterprise; these efforts are discussed in chapter 11. In 1967 NCAR served as lead agency for the Line Islands Experiment in the tropical Pacific, the first occasion in which it organized and managed a research program involving several institutions.

meeting of the Federal Council and my concern received support, to
which Hollomon responded by replacing "Meteorology" in the title by
"Meteorological Services and Supporting Research;" and with that
awkward title it was approved by the council and functioned usefully
for many years. I thought it possible that Hollomon's original objective
may have been just what I feared and that I had prevented, at the least,
meddlesome confusion, or, at the worst, centralization of all atmospheric research under the Department of Commerce.

Throughout the time I was at OST Hollomon was quietly preparing
plans to link the three commerce agencies: Weather Bureau, Coast and
Geodetic Survey, and the Central Radio Propagation Laboratory. The
planned organizational changes were not discussed openly, and I knew
little about them. Early in 1965, after I had returned to the UW, creation of the Environmental Science Services Administration (ESSA)
was announced, and Bob White was appointed as administrator of the
new agency. Creation of ESSA gave visibility to the growing importance
of environmental problems and the responsibility of the federal government to understand them. However, it did not much affect what
each of the three component agencies actually did because it provided
little increase in resources, at least initially, and because many in the
affected agencies did not welcome the step, a response Hollomon and
White undoubtedly expected. If I had been asked, I would have supported ESSA's establishment because it recognized the need for a federal agency with central responsibility for atmospheric and oceanic observations and research.

I attended monthly meetings of ICAS that were devoted largely to
collection of comprehensive, coherent data on federally supported atmospheric research. Much time was devoted to discussing terminology
and to uniform classification and accounting of research activities, and
I got my first impressions of how pervasive and parochial interagency
competition was.[33] Atmospheric research budgets of the various agencies totalled $201 million in Fiscal Year 1963 and $235 million in FY
1964 [Interdepartmental Committee for Atmospheric Sciecnes (ICAS)
1964], and future budgets were projected to rise even more sharply.
These totals were reaching levels that attracted the increasing scrutiny

---

[33] I was glad to recognize broader objectives and higher standards expressed
consistently by Joshua Holland of the AEC and Arthur Maxwell of the Office of
Naval Research (ONR).

of BoB and of Congress. I undertook to review this body of research directing attention at individual projects, results, and costs. I needed this information to do my job properly, and I hoped that the study would be useful as a report to OST. I was able to get good data from the Weather Bureau, NSF, and AEC, where procedures were in place to evaluate research and where the administrative officers knew me and I them (Bob Culnan at the Weather Bureau, Fred White and Earl Droessler at NSF, and Josh Holland at AEC); but it was harder to get satisfactory data from the other agencies. Analysis involved comparing publications, citations, collaborations, degrees earned, and costs.

The Weather Bureau's General Circulation Laboratory was being developed under Joe Smagorinsky at its first home on Pennsylvania Avenue; I was an occasional visitor and observer, and was glad to support what was a highly promising new initiative. Smagorinsky was engaged in recruiting some of the best available young theoreticians, several from Japan. Later, the laboratory, renamed the Geophysical Fluid Dynamics Laboratory (GFDL), was moved from Washington to Princeton, where it became a joint NOAA–Princeton University laboratory, and under Joe's insistence on high standards became recognized as a major international center in modeling of ocean and atmosphere phenomena. Graduates of this program have become some of today's leading geophysical fluid dynamicists.

In October Werner Baum, vice-president of the University of Miami, asked if I was interested in becoming dean of the College of Geosciences after my tour in Washington. I was not greatly interested, but I felt that uncertainty concerning the fate of the loyalty oath made it desirable for me to keep options open. I made a pleasant and instructive trip to Miami in early November, where I learned that the chief problem facing a new dean was to link effectively the new atmospheric program with the already mature and somewhat idiosyncratic marine program. I concluded that the task was not one I wanted to take on. Subsequently, after Fred Singer and Warren Wooster each held the Miami deanship for short periods, I felt that my decision had been wise.

About the middle of November Jerry Wiesner announced that he would resign at the end of 1963 to return to MIT, and Donald Hornig was chosen to replace him. This disruptive news was followed on November 22 with the assassination of the president. Images of that weekend remain as sharply etched in my mind as a book of black and white photographs: Ed Wenk standing silently looking out his window across the ellipse; the OST staff in Wiesner's office silently watching the television screen as news came in; Jerry receiving a call from George Reedy, special assistant to Lyndon Johnson; Dave Robinson commenting bitterly on right-wing incitements to violence; in the late afternoon the silent crowd in Lafayette Park facing the White House; on Monday morning the sight from the Executive Office Building, where I was attending a Bureau of the Budget hearing, of the Kennedy family and General de Gaulle walking in Pennsylvania Avenue on their way to St. Matthew's Cathedral for the funeral service; and many more.

President Johnson reaffirmed Kennedy's appointments and policies, but changes at OST were immediate and profound. Whereas Wiesner had been a personal friend of Kennedy's and had seen him alone on a regular basis, he had no such relationship with Johnson and communications with the President now had to pass through White House assistants. Although Don Hornig was well informed and experienced in Washington, when he took office in January he didn't know agency leaders as Wiesner had; and this introduced viscosity into the whole process. Most important, whereas Kennedy was fascinated by science and technology and tried to understand them, Johnson saw science and technology only as tools; and he kept scientists at arms length. So the science advisor and his staff operated with diminished authority and effectiveness.

I learned that Don Hornig had worked at Woods Hole during the war where he had gotten the lasting impression that oceanographic research being done there was mediocre compared to basic research in chemistry and physics with which he was familiar. Stemming from that experience, it was not surprising that he tended to be suspicious of the quality and importance of NCAR research. He was obviously objective, highly intelligent, and fair minded; but I felt that he looked at NCAR's program and staff with some skepticism; and this added to the significance of the in-house report I was preparing on atmospheric research.

The scientific community tended to consider OST the voice of science at the White House, while the BoB and the Congress viewed the OST responsibility as providing objective and critical evaluations of science. To the extent that these functions differed there was tension between them, and the BoB always looked for evidence of special advocacy. I tried to be objective in these matters, but my background insured that my advice inevitably was suspected of being influenced by advocacy for universities, NSF, and NCAR.

Influential senators and congressmen, dissatisfied with growth of
the oceanography program, had initiated a move to create an independent oceanographic agency, a move opposed by OST and BoB. To head
this off Ed Wenk was asked by Wiesner to stimulate oceanographic activities within the existing structure and to actively guide the ICO,
which he did very effectively. This was one of many ways in which politics impinged subtly but marginally on OST's work. During the time I
was at OST Wiesner and Hornig were able to hold Congressional pressures at bay quite effectively.

However, marine interests continued to press for an independent
agency committed to development of marine resources, a "wet-NASA,"
as it was sometimes described. A flurry of oceanographic bills introduced in Congress during the spring of 1964 proposed various forms of
reorganization and placed pressure on the Executive Office and especially on Don Hornig. In response, he appointed an ad hoc OST Panel
on Oceanography, chaired by Douglas Brooks, president of Traveller's
Research Center, on which I served, to examine the ICO interagency
program. The report was critical and led to a broader examination of
oceanographic policy by a PSAC Panel on Oceanography, chaired by
Gordon J.F. MacDonald, on which I also served after I had returned to
the UW. The panel conducted an extensive review and in July 1966 recommended reorganization that would bring together ESSA, parts of
the Bureaus of Commercial Fisheries and Mines, much of the Coast
Guard, and other smaller units (PSAC Panel on Oceanography 1966).
Also in 1966 President Johnson created the National Council on Marine Science and Engineering and appointed Ed Wenk its director. A
year later, in response to passage of the Marine Resources and Engineering Act, the president appointed the Commission on Marine Science, Engineering and Resources (Stratton Commission) to outline new
policy directions.[34] I served as a consultant to the council and the commission. The commission recommended creation of a new independent
agency to include ESSA from Commerce, the Coast Guard from Trea-
sury, the Bureau of Commercial Fisheries from Interior, and smaller
units from other agencies. The Commission paid little attention to atmospheric aspects, most of the recommended growth was in marine de-

---

[34] This sequence of actions describes, but only dimly, the chaotic seascape of
cross currents and interacting waves that characterized oceanographic policy
during this period. For a fuller account see Wenk (1972).

velopment, and the increasing importance of environmental problems was largely ignored. In this way the insight that had led to creation of ESSA was obscured and largely disregarded by the Commission.

President Nixon in 1970 responded to the Stratton Commission report by issuing an executive order that created the National Oceanic and Atmospheric Administration (NOAA) as part of the Department of Commerce, but omitted the Coast Guard from the new agency. So NOAA was launched in a subordinate administrative position lacking an operational arm, a substantially weaker agency than had been recommended by the Stratton Commission.[35] I believe that this, together with the emphasis on marine resources and corresponding de-emphasis of atmospheric objectives, has contributed to confusion as to the new agency's central mission and to many of the adversities it has suffered in later years.

While I was at OST in 1963–64 other issues of evaluation and advocacy arose. As a member of the NAS/CAS Panel on International Meteorological Cooperation I worked on a study of the feasibility of a global atmospheric experiment; and, as chair of the Sub-panel on Boundary Layer Processes, held a meeting in my OST office. I also served as chairman of the NAS/CAS Working Group on Ocean Surface Observations. In these cases I may have skated across the line between evaluation and advocacy, though I felt that I was acting within my responsibilities. Representing OST I attended a meeting of the ICSU Committee on Space Research in Florence, Italy, which helped to resolve a jurisdictional dispute between the World Meteorological Organization (WMO) and the International Council of Scientific Unions (ICSU) in planning the global experiment. And I drafted the words that provided President Johnson's endorsement of the Global Atmospheric Research Programme in a speech at Holy Cross University in the spring of 1964. This statement was reaffirmed in 1966 as the centerpiece for the president's support of World Meteorology Day; and the World Weather Watch was identified in 1967 as a continuing commitment of the President.

The NASA budget for developing meteorological satellites was far larger than other agency meteorological research budgets, but otherwise NASA supported very little atmospheric research in spite of the

---

[35] The same executive order created NOAA and the Environmental Protection Agency (EPA). No explicit linkage between the two agencies was provided for in the executive order.

fact that NASA's first objective as stated by the legislation establishing the agency is "expansion of human knowledge of phenomena in the atmosphere and space." Although NASA expended $50 million per year on development of meteorological satellites, its meteorological research support amounted to less than $2.5 million. Based on these facts, I proposed that NASA develop and support a broad program of atmospheric research; and noted that this would link research results to satellite design and development, with resulting efficiency and economy, and that the NASA program would provide expanded resources and help to establish new higher standards for atmospheric research. Deputy Administrator Hugh Dryden, who incidentally served with me as a member of NAS/CAS, replied that NASA was operating under directives stressing large, costly programs of space exploration and was committed to those objectives. NASA would do nothing toward stronger atmospheric research support without changed Congressional directives.[36] My proposal had little support outside the OST staff, so went nowhere. A year later, as an OST consultant, I tried again without success. I have wondered whether the result would have been different if my proposal had been made earlier—during the Kennedy–Wiesner regime.[37] Ten years later NASA initiated broad support for atmospheric research at universities and in NASA laboratories, and since then NASA has been responsible for much frontier atmospheric research, with a larger budget invested in atmospheric research than other federal agencies. My earlier proposals had nothing to do with this change in policy; it resulted from burgeoning concerns about possible changes in climate and stratospheric ozone together with NASA response to budget pressures.[38]

---

[36] It can be noted, however, that NASA provided leadership in satellite observation through planning done under Working Group 6 of the Committee on Space Research (COSPAR) and in numerical modeling of the general circulation at the Goddard Institute for Space Studies. And the program of NASA graduate fellowships included a few awarded to students in the atmospheric sciences.

[37] In 1969 a similar proposal focused on support for GARP was made through the State Department by Walter Roberts, president of UCAR, also without result.

[38] A 1970 summer study by the NAS Space Science Board, chaired by Herbert Friedman, in which I participated, may have contributed subtly to NASA's change in policy.

By the spring of 1964 my enthusiasm for the job had waned. The science advisor and his staff had lost effectiveness; much of the work had become routine and repetitive; and science was being eclipsed as a national priority by the Vietnam War. I had had enough of government life. When the Supreme Court decision in the loyalty oath case opened the way to my returning to the UW, I told Don Hornig that I would resign at the end of August. He asked me to continue for the following year as a consultant, and in 1965 he appointed me to the PSAC Panel on Oceanography, referred to earlier.

My in-house report on atmospheric research was completed and turned in to Don who reported to me that it had raised his assessment of NCAR research, and I believe it provided valuable facts and insights to OST and BoB. The report concluded with five recommendations of major policy significance: agencies should utilize external examiners in reviewing research, agency scientists should plan their own research, agencies should maintain accurate records of research supported by their funds, environmental research relating to effects on health should be strengthened, and NASA should take a leading role in meteorological research. Response of agency administrators was variable and limited to effects on their own agencies. Hugh Dryden responded to the NASA recommendation as he had earlier. Bob White was appreciative of the report, though he stated a number of reservations about my comments relating to the Weather Bureau. Herb Hollomon's reaction was especially caustic (though non-specific), reflecting his general hostility to my role during the time I was at OST.

On another front, following the Alaskan earthquake of 1964 Don asked me to convene an ad hoc panel of geophysicists to review the earthquake's significance and consequences. The report of this panel, chaired by Frank Press, stimulated expansion of seismological monitoring and research. During the study Gordon Little pointed out to me that ionospheric records at Boulder, Colorado, and elsewhere indicated that displacement of the earth's surface by the earthquake had created a large amplitude gravity wave in the ionosphere, a phenomenon that, so far as I know, had not been observed or predicted previously. And, working with William Hooper, OST staff specialist in business management, I completed a survey aimed at identifying common factors in the backgrounds of outstanding geophysicists that turned up some interesting relationships bearing on recruitment of graduate students, but had no tangible effects on government policy. I left Washington with enhanced understanding of

national and international science affairs, but without a case of
Potomac fever.

A few months later I was asked, as an OST consultant, for evaluation of a project proposed by NCAR. When the proposal was subsequently disapproved by the BoB, Walt Roberts protested the decision
and learned that my evaluation had influenced the outcome. He evidently felt that I had been disloyal to NCAR, and as a result I was criticized from several directions. After the efforts I had made over the previous two years to strengthen NCAR's standing with the Executive
Office of the President, I felt the criticism unjustified.

Also as a consultant, I was asked for ideas that might stimulate international cooperation, especially in southeast Asia. I suggested that
a United Nations International Center for Environmental Sciences
could play a valuable role and that the center that had functioned in
Bombay during the Indian Ocean Expedition could provide a start. Although it was reported to me that this suggestion had stirred interest
and was being examined further, I heard nothing more about it. It
seems likely that it suffered the fate of many initiatives fed into the
interagency machine, especially those of international scope. It was
to be six more years before the U.N. Environmental Programme was
launched in response to international initiatives and the stimulus of
Maurice Strong of Canada.

Chapter 6

Experiences on the International Stage

An elaborate structure of international science organizations has evolved out of the need to establish regular channels of communication among scientists, both within disciplines and across them. The International Union of Geodesy and Geophysics (IUGG) is one of many disciplinary unions that operates under the International Council of Scientific Unions (ICSU). It convenes in general meetings at four-year intervals, each attended by several thousand scientists.[39] Subordinate bodies represent the various geophysical disciplines; among them is the International Association of Meteorology and Atmospheric Sciences (IAMAS),[40] which meets biennially. These meetings afford opportunities for scientific communication and for developing continuing relationships among scientists from different countries.

These organizations also have provided international structures within which research programs have been organized and managed. For example, ICSU in 1953 formed the Comité Spécial de l'Année Geophysique Internationale (CSAGI) to plan and oversee the International Geophysical Year (IGY) that was carried out in 1957 and 1958.[41] The UW role in the IGY was described in chapter 2. Especially notable successes of the IGY were the impetus given to satellite instrumentation and observation, demonstration that ice and ocean sediment cores could reveal records of paleoclimate, and monitoring of atmospheric carbon dioxide that led to recognition of significant and continuing increase in its concentration. In the following decades ICSU assumed an important leadership role in creating the Committee on Space Re-

---

[39] Prior to 1963 the IUGG met at 3-year intervals.

[40] In 1993 the former International Association of Meteorology and Atmospheric Physics (IAMAP) was renamed the IAMAS; IAMAP met at 4-year intervals between IUGG meetings.

[41] Success of the IGY as an entrepreneurial enterprise was immortalized in the words heard often when international geophysical programs were proposed in later years, "Old scientists never die, they reinvent the IGY."

search (COSPAR) that stimulated and guided space research in its critical early years. And, as described in later chapters, ICSU played an important role in the organizational steps leading to the Global Atmospheric Research Programme (GARP) and to later programs.

The World Meteorological Organization (WMO) was established in 1947 within the United Nations to establish and maintain high and uniform standards for meteorological observations and to insure efficient transfer of meteorological data. It took the place of the International Meteorological Organization that had been formed following the First Polar Year (1888). In many research programs the WMO, representing governments, and ICSU, representing the scientific community, have cooperated effectively. However, in the early planning phase of the Global Weather Experiment and in other cases the two organizations have found themselves at odds over objectives and leadership; and special efforts have been necessary to achieve cooperation. More of this later.

My first contact with international organizations occurred in 1957 when I attended the 11th General Assembly of the IUGG in Toronto, Ontario, Canada, where, as noted earlier, I met Joost Businger. At this meeting I gave a paper on baroclinic instability and became aware of the somewhat Byzantine structure and procedures of international science organizations. The next year I was on sabbatical in England and was able to attend an international conference on boundary layer turbulence held at Queens College, Oxford. This provided opportunities to hear papers by Sir G.I. Taylor, A.M. Obukhov, and other distinguished scientists. Present also were Soviet security officers to monitor the Soviet scientists present, including Obukhov, leader of the Soviet delegation. The security people were a source of somewhat uncomfortable amusement for many of the western scientists. On one occasion Obukhov made a jocular comment at his monitors' expense in English that I and many others in the audience were unable to understand, but that conveyed an intriguing sense of the complex sociology of Soviet science.

In preparation for the 1960 Assembly of the IUGG I was enlisted to write a report on Arctic meteorology as part of the U.S. National Report. At this meeting, held in Helsinki, I gave a paper on an aspect of air–sea interaction that I have forgotten. For the 13th General Assembly, held in Berkeley, California, I prepared a report on air–sea interaction as part of the U.S. National Report to the IUGG and served as convenor for a Symposium on Interaction of the Atmosphere and Ocean.

ICSU and WMO played essential roles in steps that led to planning and carrying out the Global Atmospheric Research Programme, and later the World Climate Research Programme, and the International Geosphere–Biosphere Programme. These steps began with United Nations Resolutions 1721 (1961) and 1802 (1962) inviting the WMO and ICSU to develop a weather research program, whose origins were described in chapter 4.[42] The WMO responded promptly by proposing the World Weather Watch (WWW) to be carried out through government weather services, and it created the WMO Advisory Committee with broad responsibility for research and use of satellite data. The Advisory Committee was regarded by influential scientists as infringing on their responsibility for research, and there appeared to be danger that the WWW and the international research program might remain separate and uncoordinated, with the result that they might become competitors for resources. These issues were placed on the agenda of the COSPAR meeting held in Florence, Italy, in 1964. As noted in chapter 5, I attended as a staff member of the Office of Science and Technology that was concerned that the observational and research components move forward as a unified program. Largely through the efforts of Tom Malone, U.S. representative to ICSU, agreement was reached on appointment of the ICSU Committee on Atmospheric Sciences with responsibilities for planning the science program and coordination with the WWW. Later, this committee and the WMO committee were folded into the new Joint Organizing Committee (JOC), chaired by Bert Bolin of the University of Stockholm. Beginning in early 1968 and continuing over the following decade JOC studies and its decisions proved to be crucial in planning and execution of the joint program.

In August 1971 I attended the 15th General Assembly of the IUGG held in Moscow where I gave a paper comparing independent measurements of boundary layer fluxes, especially memorable to me because I had to contend with the aftermath of food poisoning, suffered the previous night. My strongest memories and impressions of the visit spring from the cultural shock that I and my wife, Marianne, experienced in encountering officials, organizations, and citizens of the Soviet Union over the two weeks we were there. Upon arrival at the Moscow

---

[42] The U.N. resolutions were responses to President Kennedy's 1961 speech to the U.N. in which he proposed "cooperative efforts in weather prediction and eventually in weather control," described earlier in chapter 4.

airport my name was not found on the list of U.S. delegates, so we were consigned to a guarded holding area for 4 to 6 hours, until ultimately released by an unknown process. Halls and stairways at the University of Moscow, where meetings were held, were heavily decorated with anti-U.S. graffiti, some aimed sharply at race prejudice, gun violence, and the gap between rich and poor, but much of it exaggerated and far off the mark. On the other hand, individual citizens were usually understanding and friendly and often expressed tearful hopes for international peace. But every contact with individuals in charge of something, whether government officers, hotel personnel, or shop clerks became problematic and stressful.

Marianne had been commissioned by a Seattle friend to contact a relative in Moscow and to deliver some items to her. Marianne was surprised when, after phoning her from a public phone to avoid the call being traced the relative insisted on coming to our hotel room where she expressed critical political opinions quite openly.

Joost Businger, who also attended the meeting, and I had been invited to visit the offices of the Mir Press, which had printed a Russian translation of *An Introduction to Atmospheric Physics*—without recognition of the Academic Press copyright. The editor and several officers entertained us in the editorial office where we were served tea. After polite conversation we were each presented an envelope containing half of the royalties that had accumulated from sale of the book, in rubles that could not be taken out of the country. As I recall each of us received between 300 and 400 rubles.

Joost and I and our wives had arranged to drive a rented car from Moscow to Leningrad, stopping overnight in Novgorod. We drove first through an area of simple, attractive dachas and then through extensive communal farms. On the first day we stopped for lunch at a large cafeteria serving agricultural workers. Many were drunk, and some resented our presence and made the lunch an unpleasant experience. In Novgorod we found our Intourist guide and discovered that the Busingers had no hotel reservation. After two hours or so calling Moscow, the guide was able to find them a large suite that included a piano and television set. *The Forsythe Saga* was being shown in English on Soviet television.

Leningrad provided respite from the stresses we had been experiencing since arriving in Moscow as well as fascinating visits to the Hermitage and the Summer Palace. At the Sadko restaurant we encountered a light-hearted international clientele and the best food and drink since arriving in the USSR. We returned to Moscow by train and noted that train workers, including security guards and those working on track repairs, were women. Back in Moscow, in an effort to spend more of our rubles I traveled several miles to a highly recommended restaurant to obtain a written reservation for the following evening. Upon arriving at the restaurant we found a large crowd milling about the closed entrance. After a few minutes the door opened briefly and one or two people slipped in. I had to fight our way to the door waving my reservation, where we finally were able to squeeze through into a large restaurant in which only a few customers were distributed among the tables. We were not able to solve the mystery; we assumed it had something to do with either an effort to minimize effort of the staff or a limited food supply. The food was only somewhat better than our hotel fare.

We learned that our reservations for a postassembly trip to Tashkent had been canceled due to delay in construction of a new hotel. We could apply for a refund from a travel agent in Boston; ultimately the refund amounted to, as I remember, only a little more than half of what we had paid. We were able to change our plane reservations and to leave Moscow at the end of the assembly and felt relieved when no new Kafkaesque difficulties arose in our passage through the airport. On finally entering the departure lounge for London we were greeted by a British Airways attendant whose cheer and efficient helpfulness made us feel we had finally reentered a bright and shining new world.

In the early 1980s I served a term as a member of the U.S. National Committee to IUGG and attended the 1981 meeting of IAMAP in Hamburg, Germany, and the 18th General Assembly of IUGG in 1983, also in Hamburg. At the latter meeting, as a delegate representing IAMAP, I successfully proposed that IUGG endorse Canadian efforts to establish a ship of opportunity observational program in the North Pacific, as replacement for the terminated Weather Ship P. I attended the IAMAP meeting in Honolulu in 1985, where I gave two papers reporting research results from the Storm Transfer and Response Experiment (STREX), and in 1987 I attended the IUGG General Assembly in Vancouver, British Columbia. A few other international experiences are described in later chapters.

The Global Weather Experiment

The NAS/CAS study of the feasibility of a global atmospheric experiment was completed in 1965 and was reviewed at a PSAC meeting early in 1966 (NAS/CAS 1966). Jule Charney, chairman of the panel that had conducted the study, summarized the report; and I attended the meeting. Charney began in his usual rambling style, giving the impression that he might be thinking about the topic for the first time. PSAC regarded meeting time as a precious commodity. I noticed members exchanging annoyed glances and muttered comments. At other PSAC meetings that I had attended similar reactions had led to disapproval of proposed projects, so prospects for the global experiment did not look good. But within a few minutes, as Jule gathered his thoughts together, he gradually got the attention of members of the committee and challenged them with the excitement of mounting a test of the global atmospheric circulation. The meeting ended with PSAC approval to proceed to the detailed planning stage for the global experiment.

Detailed plans were drawn up in 1968 and 1969 by Dick Reed working under the direction of the officers of the U.S. Global Atmospheric Research Program Committee: Jule Charney, Joseph Smagorinsky, and Verner Suomi (U.S. Committee for GARP 1969). These plans represented an evolution in concept from a single grand observation period of about one month, as originally visualized, to a series of research projects extending over a period of years that would be based on a developing global observing system and would focus on different specific objectives. This plan provided major substance for the ICSU–WMO Planning Conference, held in Brussels, Belgium, March 16–20, 1970. The conference brought together scientists, officials of national weather services, and officers of ICSU and WMO. Bob White was the U.S. official delegate and the key participant. I attended as one of several U.S. delegates whose chief function was to advise Bob White, a largely superfluous function. International protocol led to procedures that were ponderous and political, and the contrasting views of ICSU and WMO sometimes led to frustrating difficulties. However, enthusiasm for the research objectives was high; and the conference ended with approval for moving ahead to planning for tropical experiments and the first global experiment. The world's weather services and scientific organizations were finally reading from the same page, some nine years after the proposal for launching international observational and research programs. Prior to the Brussels conference it seemed necessary to identify the World Weather Watch and GARP as separate in concept and management; after Brussels they could be linked more closely to the Global Weather Experiment and to the objective of improving weather forecasts.

Initial plans for an air–sea interaction experiment had identified the tropical Pacific Ocean as the preferred site, and the Line Islands Experiment of 1967 had been designed and carried out there by NCAR with university participation, in part as a pilot project. However, discussion of an international project involving ships and aircraft ran afoul of Soviet concerns about possible spying on its intercontinental ballistic missile experiments. Also, the logistics for an experiment in the Atlantic Ocean was far simpler; and the Soviet Union, East Germany, and West Germany were willing to provide ships only for an experiment in the Atlantic. The Barbados Oceanographic and Meteorological Experiment (BOMEX), for which planning by the United States was already far advanced, was carried out in 1969. Joshua Holland, who had gotten his Ph.D. at the UW, served as scientific director. I was a principal investigator for a BOMEX project and served as chairman of the BOMEX Advisory Committee. The more complex GARP Atlantic Tropical Experiment (GATE) was planned for the eastern Atlantic and carried out there in 1974. Dick Reed served as consulting scientist for GATE, and Bob Houze from the UW was a major participant, as was Peggy Lemone, an NCAR scientist who had been a former student of mine. This experiment and other regional projects led to the Global Weather Experiment carried out from December 1978 to December 1979. I was not a participant in these projects, but I chaired a panel that prepared a report on educational implications of the GARP (Fleagle 1970).

The regional experiments and the Global Weather Experiment were focused on the first GARP objective of "increasing the accuracy of forecasting over periods from one day to several weeks"; they profoundly influenced meteorological research throughout the 1980s and resulted in substantial improvements in weather forecasting. Research horizons were extended well beyond those recognized during the early development of the World Weather Watch and GARP.

CHAPTER 8

Expansion of the Department

During the time I had been in Washington in 1963 and 1964 Norbert Untersteiner had returned from Austria to serve as chief scientist for Project ARLIS, a continuation of UW research based on an Arctic Ocean field station. Norbert was also appointed to the academic faculty to develop graduate study in glaciology as part of the department and geophysics programs. His first Ph.D. student, Gary Maykut, completed his degree in 1969 and since then has been a stalwart pillar of the glaciological research program. Norbert's subsequent career has included many advisory and administrative posts in the federal government and in international glaciological organizations. In 1992 he was elected to the Austrian Academy of Sciences. At the UW he has served over the years as director of the Polar Science Center, interim dean of the College of Ocean and Fisheries Sciences, first director of the Joint Institute for Study of the Atmosphere and Ocean (JISAO), and from 1988 to 1997 as chairman of atmospheric sciences. Upon retirement from the UW he was named the Sydney Chapman Professor at the University of Alaska.

In 1963 Peter Hobbs, on John Mason's recommendation, joined the faculty to develop a program in cloud physics, the first of the three faculty appointments I had negotiated with Dean Katz, as noted in chapter 4. With his characteristic concentration and industry Peter recruited several physics students and set about developing courses and a cloud physics research program combining laboratory and field observations. The first Ph.D. in cloud physics was awarded to James Dye in 1967, and four other cloud physics degrees were awarded the following year. In 1969 a B-23 aircraft was acquired to extend research to include properties of water and ice in clouds. The B-23 and its successor aircraft became crucial parts of the national meteorological research fleet, carrying out critical observations in natural clouds and also in volcanic plumes, dust clouds, and in smoke from fires following the Gulf War.[43] Peter's Cloud and Aerosol Research Group includes numerous

---

[43] Since 1996 the Group has operated a well-instrumented Convair 580.

scientists and technicians and represents a major component of the department research program. Peter Hobbs is the author of the definitive book, *Ice Physics*, and of *Physical Chemistry for the Atmospheric Sciences*, as well as many research papers and monographs on the physics and physical chemistry of cloud particles and aerosols. He received the AMS Charney Award in 1984.

James Holton was recommended by Jule Charney to fill the dynamic meteorology opening, and after a year as NSF post-doctoral fellow at the University of Stockholm, he joined our faculty in 1965. The first student to earn the Ph.D. degree under his supervision was Walter Meyer in 1969. Jim's research has been especially important in clarifying the theory of mesosphere structure and circulation and of the quasi-biennial oscillation. His book on dynamic meteorology has become the standard upper division text on this subject. Jim was awarded the AMS Second Half Century Award in 1982 and its Rossby Medal in 2001. He was elected a member of the National Academy of Sciences in 1994. He has served as chair of the department since 1997.

In 1966 John Michael Wallace was appointed to a new position in synoptic meteorology, largely on the basis of strong recommendations by the MIT faculty and by Dick Reed. His first Ph.D. student was Vernon Kousky who received his degree in 1969. Mike's research has been instrumental in developing an essentially new field, the use of empirical orthogonal functions to reveal teleconnections relating meteorological phenomena in different regions of the earth. As director and codirector of the Joint Institute for Study of the Atmosphere and Ocean (JISAO) from 1978 to 2000 he built it into an active research and study center and an important component of the department.[44] In 1977 he and Peter Hobbs published the very successful text, *Atmospheric Science: An Introductory Survey*. Mike served as chair of the department from 1983 to 1988. He received the Rossby Medal in 1993, the highest honor awarded by the AMS, and was elected a member of the National Academy of Sciences in 1997.

Masahisa Sugiura, whose previous work was in solar and magnetospheric physics, was appointed to a faculty position in 1966 to develop an aeronomy program that would link effectively to the rest of the de-

---

[44] Establishment of JISAO in 1977 as a joint institute of the UW and the National Oceanic and Atmospheric Administration (NOAA) is discussed in chapter 9.

partment. Unfortunately, he had little interest in developing the broad program that we had visualized in recruiting him. After a year he returned to the Goddard Space Flight Center (GSFC) and later took a university position in Japan. This effort to further develop a broad, coherent department by linking meteorology with aeronomy had failed.

By the mid-1960s the department faculty and the number of graduate students had nearly doubled in the preceding five years; we had occupied a second "temporary" building (the present Cunningham Hall, built for the Alaska Yukon Exposition of 1909), but were increasingly pressed for more suitable space. The department had been listed for several biennia on the university schedule for a new building. In 1965, or perhaps 1966, the University Building Committee scheduled a new building for atmospheric sciences for the next biennium; and we secured a matching NSF grant for a building to house Atmospheric Sciences and Geophysics.[45] An architectural firm was employed, and a faculty planning committee under Frank Badgley began work on specifications and space allocation. A crisis arose in early 1967 when the university building schedule was threatened with delay at a time when, as a consequence of the Vietnam War, state building funds were in doubt and the NSF budget was under pressure. I feared that if construction were delayed by another year we were likely to lose the NSF matching grant. At a meeting with Provost Katz and the University Building Committee, as chairman-designate I argued this case and was relieved when a decision was reached to maintain the original schedule. In the following year the importance of this decision was made clear when NSF building grants were terminated, or at least severely cut back.

A different crisis arose when initial architectural drawings showed the building without windows, but with vertical, randomly spaced narrow slits; the faculty quickly identified the proposed design as the "IBM

---

[45] The NSF grant specified that the building would be a science laboratory and excluded design of lecture rooms and prohibited remodeling to provide lecture rooms for five years. The synoptic laboratories were usable as lecture rooms, and I insisted that a library be included that could be converted to a lecture room when, I anticipated, a geophysical sciences library would be established in Johnson Hall. In 1984 much of the library was moved to the Quaternary Research Center Library in Johnson, and some of it is now housed in the Allen Library. The former library became the present third floor class and conference room.

card." Our objections led to plans showing a monstrous carbuncle of a window on the sixth floor, plans that we also rejected. These experiences placed on Frank Badgley and his committee unusual responsibility for educating the architects in design of a university science building. In view of these first attempts, we were relieved by the final plans. The building was occupied in the fall of 1969, and a symposium was held on November 21 to dedicate the building. Speeches were given by John R. Hogness (UW executive vice-president), Thomas O. Jones (NSF deputy assistant director), Robert M. White (administrator of ESSA), and me. I discussed education in the geophysical sciences in relation to society's interest in addressing long-term problems.[46]

When I became chairman in 1967 the department could be described as strong and on the way to becoming stronger. Research grants and graduate student applications were increasing; academic standards were in good shape for graduate students, though perhaps not for undergraduates. There were 10 academic faculty, about the same number of research faculty, and about 50 graduate students. The first task I faced in the fall of 1967 was to replace Sugiura by a specialist in physics of the upper atmosphere with broad interests in meteorology. Norbert Untersteiner proposed Conway Leovy, then at the RAND Corporation, as excellently qualified. After reviewing his record and recommendations, the faculty endorsed an offer, and I sought Dean Phil Cartwright's and Provost Sol Katz's approval for an offer of an associate professorship.

In December 1967 Conway Leovy joined the department on a visiting basis and became an Associate Professor of Atmospheric Sciences and Geophysics in the following September. Leovy's first Ph.D. student was Paul Try, who earned his degree in 1972. Conway has held leadership positions in investigations of the atmospheres of Mars, Venus, and of some of the more remote planets and their moons. In his teaching and in mentoring of graduate students Conway demonstrates interests and capabilities extending over essentially all fields represented in the department and related fields in other departments. In this way he is a unique resource to the department and the university. He served as director of the Institute for Environmental Studies from 1986 to 1989

---

[46] The dedication was scheduled to coincide with a Symposium on Early Results from BOMEX. A summary of the BOMEX symposium and the papers read at the dedication were published in Fleagle (1970).

and has helped to organize and has taught in interdisciplinary courses involving political scientists, physicists, biologists, oceanographers, and astronomers. He was awarded the Year 2000 Gerard P. Kuiper Prize by the American Astronomical Society and was chosen as the Bernhard Haurwitz Lecturer by the AMS in 2000.

At the outset of my chairmanship I felt that my most important responsibilities were to insure appointment of outstanding faculty and to support initiatives by productive faculty, and otherwise to stay out of the way. I also believed that we needed stronger support for administrative and technical functions and was able to establish new staff positions for a manager of technical services and a manager of administrative services. My experience at OST had convinced me that the atmospheric sciences were in the central position to address environmental problems that were assuming increasing national and international importance, but that, to do this, there was need for greater recognition of high quality research being done by atmospheric scientists, especially recognition by national leaders in the basic sciences. On the local level, I believed the department should develop stronger relationships with other departments and institutions on and off the campus. This led me to a number of efforts, not all successful.

Under the Graduate School of Public Affairs I organized an interdisciplinary series of 11 seminars on weather modification that included national leaders in the field and was published as part of the Natural Resources Public Policy Seminar Series (Fleagle 1968). I helped to organize a group of faculty from several departments that produced a report proposing unification of the marine units and programs on the campus. This was the first of several steps that led eventually to creation of the College of Ocean and Fisheries Sciences and within it the Institute for Marine Studies. Under the leadership of Ed Wenk, who had moved from government to the UW in 1970, I participated in an interdisciplinary research project addressing how society might handle new technologies. As one of five subprojects, my study of weather modification included discussion of major programs, impacts, decision-making, and management alternatives (Fleagle et al. 1974). When the Institute for Environmental Studies was created I chaired the committee to select the director; we chose two candidates whose vision and capabilities were endorsed by the committee; but both were passed over by the administration, with the result, I believe, that the institute operated for 25 years within a shadow from which it was not able to escape.

In the early 1970s Physics Professor J. Gregory Dash organized a faculty seminar on Conflict Resolution that I joined. This led to several faculty efforts to develop interdepartmental curricula relating to objectives of peace, justice, and conflict resolution. One of these led to an interdepartmental course on consequences of nuclear war in which I participated. Later, through Greg Dash's efforts a Science Advisory Committee was formed to provide advice on science issues to Congressman Mike Lowry and occasionally to Senator Brock Adams. Members were Dash, George Wallerstein (Astronomy), Arthur Kruckeberg (Botany), David Stadler (Genetics), Gene Woodruff (Nuclear Engineering), me, and for part of the time Pete Rose (independent physicist). The committee met weekly for several years. We prepared briefing materials on Hanford's problems of waste disposal and employment, biodiversity, environmental problems of the Northwest, NOAA's potential importance to the region, and perhaps other topics I have forgotten. In our occasional meetings with Lowry and Adams they seemed to have read our reports and were generally well informed on the discussion topics; so we concluded that we had performed a useful educational function. In February 1988 Lowry, responding to our recommendations, introduced a bill to make NOAA an independent agency; but the bill did not make it through the legislative mill, nor did our other efforts lead to completed legislation or to visible changes in government policy.

Several needs and opportunities for faculty appointments arose in the late 1960s and early 1970s. To complement Peter Hobbs's rapidly growing research in microphysics of clouds, Alistair Fraser, whose interests were in mesoscale structure and observations of clouds, was appointed in 1969. After three years he left to join the faculty at The Pennsylvania State University, where he has become known for his interpretations of the observations of clouds and optical phenomena. Robert Houze, whose thesis at MIT was based on analysis of radar cloud observations, replaced Fraser in 1972. His research initially was supported by Hobbs's funding, but he soon obtained his own research support. Houze's first Ph.D. student was Colleen Leary, who was awarded her degree in 1978. Bob was principal investigator of major programs as parts of GATE in 1974 and the Monsoon Experiment in 1978 and 1979, and has developed a flourishing research group. He published a major text titled *Cloud Dynamics* in 1993.

In 1970 or 1971 the department prepared a Six-Year Plan that emphasized the following as high priority needs: a) new faculty positions in atmospheric chemistry and atmospheric radiation, b) replacements for Konrad Buettner, who had died in 1970, and for Phil Church, who was retiring in 1972, and c) support for upgrading of our data processing capabilities. On the basis of the Six-Year Plan the UW administration selected atmospheric sciences as one of several departments to submit proposals to NSF under the University Science Development Program. We were awarded a grant providing two years of support for two new faculty positions, a programmer, and related facilities. The grant was intended to provide start-up money for faculty positions in atmospheric chemistry and radiation or remote sensing that would be picked up by the university at the end of the grant period.

Halstead Harrison had been a productive research chemist at the Boeing Scientific Research Laboratory (BSRL); when BSRL closed during one of Boeing's financial crises, he applied for our atmospheric chemistry position. He had worked on the stratospheric ozone problem and had shown integrity and courage in calling attention to possible threats to the ozone layer from a fleet of supersonic planes. He was appointed to the science development position in 1972; a year later when it was evident that the university budget was not able to pick up the position Harrison was appointed to the position formerly held by Buettner. Harrison's first Ph.D. student was Tim Larsen, who earned his degree in 1977. Harrison has taught graduate courses in atmospheric chemistry and in modeling of aerosols and has served on the Faculty Senate Executive Committee. His research has been published largely in chemistry journals, and often may be overlooked by atmospheric scientists.

Kuo-nan Liou, who had received his Ph.D. at Johns Hopkins, was appointed to the science development position in atmospheric radiation. He worked productively on research and taught several courses, but at the end of his two-year appointment a state budget cut prevented creation of a new state-supported position. He was appointed to a faculty position at the University of Utah. After achieving a distinguished record there and publishing a major text on atmospheric radiation, he moved to UCLA several years ago.

Robert Charlson, after completing his Ph.D. in 1964 under Buettner, was appointed to a faculty position in Civil Engineering where he developed a program in atmospheric chemistry. He joined the Department of Atmospheric Sciences in 1971 as adjunct associate professor; in that role his first Ph.D. student was David Covert who received his Ph.D. degree in 1974. In 1984 Charlson became Professor of Civil Engineering, Environmental Studies, and Atmospheric Sciences; he has had a highly successful interdisciplinary career and has collaborated successfully with scientists in other countries. He received an honorary doctorate from the University of Stockholm in 1993 and was appointed the King Gustav Professor of Environmental Science (Sweden) for 1999–2000.

In 1970 on Joost Businger's recommendation James Tillman joined the Energy Transfer Group (boundary layer research) as a specialist in humidity instrumentation. Later he became a member of the research faculty and a member of the NASA science team working on the Viking Program (1976 Mars lander) and on analysis of Viking data. More recently Jim has extended his scientific work by directing substantial efforts toward precollege science education for grades K–12, efforts that have been widely recognized and appreciated.

Edward LaChappelle, beginning with research on the Blue Glacier as part of the IGY in 1957–58 and continuing through subsequent years, developed an active research and teaching program utilizing observations from the Blue Glacier on Mount Olympus, and from sites in Utah and Alberta. In 1973 he was appointed to a joint academic position in Geophysics and Atmospheric Sciences. LaChappelle's first Atmospheric Sciences graduate students were Terry Fox and Mark Moore who earned M.S. degrees in 1973 and 1975, respectively.

Peter Webster was appointed in 1973 to the position in climatology made available by Phil Church's retirement. Peter spent a stimulating and productive four years with us and supervised the Ph.D. work of William Lau (1977). Unfortunately, he became impatient with the deliberate pace of university promotions to tenure. Although his promotion ultimately was approved, he accepted an offer from Penn State, an affront to the departmental ego. He has since forged a distinguished career at Penn State and Colorado State University.

To replace Webster we conducted a national search that led us to Dennis Hartmann, then at NCAR after getting his Ph.D. at Princeton in 1975. Dennis accepted our offer and joined the department in 1977. His first Ph.D. student was Starley Thompson, who was awarded his degree in 1983. Dennis has been active as a principal investigator in a series of NASA earth radiation programs and serves on a committee advising NASA on earth radiation missions. He leads the UW team investigating influences of clouds and water vapor on the net radiation balance. In 1994 Hartmann published the much needed textbook *Global Physical Climatology*, and a second edition is approaching completion.

Kristina Katsaros, who had gotten her Ph.D. in 1969 under Buettner, developed an active half-time research program focusing on boundary layer and satellite observations over the ocean. Beginning in 1977 she and Bob Brown were appointed to joint half-time academic positions in addition to each holding research positions. Brown's chief activities are summarized briefly in chapter 4. Kristina's first Ph.D. student was Gerald Geernaert who received his degree in 1983. She has gone on to administrative positions, first in 1991 as director of the Department d'Océanographie Spatiale, IFREMER-Centre de Brest, France, and since 1996 as director of the NOAA Atlantic Oceanographic and Meteorological Laboratory in Miami, Florida. Kristina Katsaros received the AMS Sverdrup Medal in 1997 and was elected to the National Academy of Engineering in 2001. The appointments of Hartmann, Katsaros, and Brown were the last to the academic faculty during my chairmanship of the department.

Subsequent evolution of the department has continued at roughly the same rate. However, it seems appropriate to end my account of the department's expansion at this point because I did not have the major responsibility for later faculty appointments. This has the unfortunate effect that achievements of the following senior members of the academic faculty are not described: Marcia Baker, David Battisti, Christopher Bretherton, Dale Durran, Clifford Mass, Peter Rhines, Edward Sarachik, Stephen Warren; and of senior members of the research faculty: David Covert, Thomas Grenfell, Dean Hegg, and Gary Maykut. Their careers provide abundant evidence that the department's trajectory remains ascendant. And the achievements of junior faculty (identified in appendix B) are still largely in the future.

During the period of my chairmanship graduate training in the atmospheric sciences became more specialized, while at the same time need for greater emphasis on interdisciplinary research became recognized more widely. Patterns of post-Ph.D. employment shifted away from the academy toward government and industry. Careers in science became more closely linked to issues of public policy. These changes occurred in many scientific fields, leading to changes in graduate programs, especially within the geophysical sciences. Since my chairmanship these trends have continued at an increasing rate.

CHAPTER 9

UW–NOAA Relationships

When the National Oceanic and Atmospheric Administration (NOAA) was created in 1970 I hoped that the Pacific Ocean Laboratory could be developed into a stronger unit that would work together with the Departments of Atmospheric Sciences and Oceanography, especially strengthening our capabilities for observational research. These ambitions were discussed with Bob White, who was receptive; and I recommended that Joshua Holland, who had gotten his Ph.D. under Frank Badgley in 1968 and had been scientific director of the Barbados Oceanographic and Meteorological Experiment (BOMEX), be appointed to direct a NOAA laboratory focusing on problems of atmosphere–ocean interaction. Josh was offered the position, but due to family circumstances felt he could not leave the Washington, D.C., area. That was a major setback to my hopes to form a stronger permanent link with NOAA, one of my biggest disappointments as chair of the department. When disposition of the Sand Point Naval Station was being considered, I testified at a public hearing in favor of establishing a major NOAA center and laboratory to collaborate with the UW. When the Pacific Marine Environmental Laboratory was being created, I recommended Norbert Untersteiner and later D. James Baker, then a research professor of oceanography, for appointment as the first PMEL director; but the appointment went to John Apel, whom I did not know at the time. Still later, in 1975 or 1976, Jim Baker and I together proposed creation of a joint UW–NOAA research unit similar to the University of Colorado–NOAA Cooperative Institute. The proposal was endorsed by the UW administration and by NOAA, and the Joint Institute for Study of the Atmosphere and Ocean (JISAO) was established in 1977 with Norbert Untersteiner as director. A year later Norbert accepted an appointment with the Office of Naval Research (ONR), and Mike Wallace was appointed as director of JISAO where, as noted in chapter 8, he developed it into a strong research unit and a major contributor to the department's program. So the objective of a permanent UW–NOAA linkage was achieved through a structure somewhat different from that originally visualized.

In the early 1970s, as environmental problems became matters of increasing public concern, I offered a course in Atmospheric Science Policy that discussed the roles of the agencies supporting research and of the NAS and its National Research Council (NRC) in advising government. Enrollments were small because most graduate students felt that they had to concentrate intensively on preparation for the qualifying exam and then on the thesis. I felt that an experimental program focusing on atmospheric science policy should be tried, perhaps under the UW Program of Social Management of Technology, headed by Ed Wenk; and I discussed this with Bob White, as NOAA administrator. Out of this came support for a small group of applicants with atmospheric science backgrounds who enrolled in 1976 for the two-year program. They took courses in economics, political science, social management of technology, and atmospheric science policy, and wrote a paper on a topic of their choice. Three students completed the program, one of whom went on to a series of policy positions in the federal government. I felt that results of the experiment were disappointing and did not propose repeating it. However, I believe that there continues to be a need in federal and state agencies for atmospheric scientists with additional training in science policy.

I also discussed with Bob White spending a year in Washington as a half-time consultant to the NOAA administrator reviewing environmental policy issues facing the agency. This was implemented through a UW–NOAA contract providing half salary plus support for a symposium on atmospheric science policy. I resigned as chair of the department, and the UW granted me a sabbatical for the year 1977–78 to spend half my time on science policy and half in working on the second edition of *An Introduction to Atmospheric Physics*. These plans encountered a sharp curve when the new Carter administration replaced White with Richard Frank whom I had never met. During the nine months I spent in Washington it was necessary for me to initiate each contact with Frank, and I charted my own path in reviewing NOAA procedures and structure. Fortunately, many agency officials provided information whenever asked and were helpful in other ways. My first sustained discussion with Frank occurred when I briefed him on my report in May 1978. I emphasized the value to NOAA of stronger links with the universities through a variety of mechanisms including research grants, short-term appointments of faculty, and NOAA–university centers or laboratories. He seemed sympathetic and understanding of what I had to say. Subsequently, NOAA–university centers were expanded and strengthened significantly and more university research support was provided. These actions may have been responses to my report, or they may have reflected George Benton's influence as associate administrator, or both.

The Symposium on Atmospheric Science Policy, financed by my NOAA grant, was held at the headquarters of the AMS in Boston in May 1978. It brought together 13 senior policy officials covering a wide range of experience and perspective for two days of intensive exchange of views. Participants recognized that demands on the atmospheric sciences will increase as environmental margins are eroded by growth of world population and by industrialization, and they emphasized that this requires closer collaboration of scientists and policy makers and the development of institutions responsive both to national objectives and to results of scientific research (Fleagle and Wolff 1979).

On several occasions during the 1970s and 1980s I served on panels or committees evaluating various parts of NOAA; these evaluations were reported through National Research Council bodies, the National Advisory Committee on Oceans and Atmosphere (NACOA), or directly to the NOAA administrators who served during those years. I gained a fairly comprehensive view of NOAA's potential and its achievements. I believed that the nation needed a flagship for environmental research and monitoring; NOAA was the appropriate agency to serve this essential role, but was falling short for several identifiable reasons.[47]

With the beginning of the Reagan administration I had more limited opportunities to influence policy directly; at the same time I felt it more important to express my views fully. I criticized as sharply as I could the efforts of the administration to dismantle NOAA in the years from 1981 to 1984. I was invited to testify on the NOAA budget by the Senate Committee on Commerce, Science, and Transportation on March 14, 1983 and criticized administration plans for the sale of weather satellites, reduction of Weather Service functions to statutory requirements, assignment of research aircraft to NOAA headquarters, and centralization of Weather Service computers under the Department of Commerce. Some of these ideas were expressed also in evaluation reports, and they were described more explicitly and coherently in papers published in 1986 and 1987 (Fleagle 1986,1987). A letter on the administration's handling of the NOAA budget published in *Science* at-

---

[47] These conclusions were summarized in Fleagle (1986).

tracted some attention (Fleagle 1984). I had testified on the NOAA budget at hearings of the House Committee on the Budget (1978) and also testified before the U.S. Senate Committee on Commerce, Science, and Transportation (1987). These activities were only indirectly related to NOAA's relationship to the UW, but I felt that the department needed a strong NOAA in order to flourish. In retrospect, I believe my efforts on NOAA's behalf may have helped Congress to defend the NOAA budget and program during a long period of stress; among other things they helped to preserve JISAO and other university centers and institutes.[48] Of course others also came to NOAA's defense. Bob White selected his interventions carefully and was especially effective; and, as described more fully in chapter 12, the AMS expressed its views in several official statements adopted by the council.

In the late 1970s Jim Baker, Mike Miyake, and I had talked about mapping the distribution of surface fluxes of heat, water vapor, and momentum in Gulf of Alaska storms. The concept was based on observing the fluxes as storms moved past the Canadian weather ship P at 50°N, 145°W in the Gulf of Alaska and compositing the results. The study, called the Storm Transfer and Response Experiment (STREX), was carried out as a joint U.S.–Canadian program in the fall of 1980. John Apel, director of PMEL, and I served as codirectors. Mike Miyake, as executive scientist, was chiefly responsible for assembling and coordinating operation of three aircraft (NOAA P-3, NCAR Electra, and NASA C-130) and the NOAA ship Oceanographer that played vital roles. STREX provided data for a number of special studies carried out by Canadian and U.S. investigators and for theses completed by my last two Ph.D. students (Wendell Nuss and Nick Bond) and for the last scientific research in which I actively participated.

From 1978 to 1982 I served on the College Council, the body responsible for advising the dean of arts and sciences on promotions and granting of tenure. It was a daunting task to apply fair and uniform

---

[48] Congress is in large part isolated from science and issues of science policy, and this constitutes a constraint on all programs of science and technology, and on evolution of the atmospheric sciences. The situation was made far worse in 1995 by action of the Republican Congress in abolishing the Congressional Office of Technology Assessment. This is not to suggest that Congress should be involved in planning and management of science programs, it should not, but that Congressional budgeting and oversight require more understanding than is provided in most Congressional hearings of science-related subjects.

standards to departments as varied as art, atmospheric sciences, chemistry, genetics, music, philosophy, and many more. For most of my time on the council Ernest Henley served as dean; he insisted on high standards and on comparable qualifications for faculty from the various disciplines. We read at least parts of each candidate's publications, spent a lot of time in discussion and reviewing documentation, and of course relied heavily on the council members most familiar with the candidate's field. In my first year on the council D. James Baker, research professor in oceanography, was proposed by a faculty selection committee as professor and chair of the oceanography department. Council members had serious reservations about approving an appointment of a candidate who had not moved through the normal sequence of academic positions prior to being proposed as professor and chairman. I believed that Baker was highly qualified and took on the task of convincing the council that the appointment should be approved. Although some of the members were not easily convinced, the appointment was ultimately approved. Later, after the election of President Clinton, I proposed Baker's appointment as NOAA administrator in letters to Commerce Secretary Daley, Senators Gorton and Murray, and the new President's Science Advisor, John H. Gibbons. I later learned that my nomination had been the first for Jim Baker, who was appointed to the post and served from 1993 to 2001.

As a member of the Graduate School Council in the late 1980s, I was involved in periodic reviews of departmental graduate programs across the spectrum of colleges, excluding medical colleges, exercises that gave me new insights into a number of departments around the campus.

CHAPTER 10

National Academy Activities

As noted earlier in chapter 4, I became a member of the joint NAS/CAS–NASCO Panel on Air–Sea Interaction in 1959; and I was appointed to NAS/CAS in 1961. The strongest focus of NAS/CAS during much of the 1960s was on preparations for GARP; when this program began to take shape in the late 1960s the U.S. GARP Committee was created to oversee the program. Other NAS/CAS activities included studies of remote sensing and weather modification. The Panel on Remote Atmospheric Probing, chaired by Dave Atlas, produced a two-volume report that reviewed achievements and recommended development of acoustic, lidar, and microwave instrumentation for making atmospheric soundings from the earth and satellites.[49] The Panel on Weather Modification published reports in 1962 and 1966 that emphasized the need for research to provide the basic understanding of processes occurring in clouds. The second of these reports included consideration of the possible effects of water vapor that would be introduced into the stratosphere by 400 supersonic transport planes. Based largely on analysis by James E. McDonald, the study suggested that the water vapor introduced into the stratosphere by burning of jet fuel might result in reducing stratospheric ozone concentration with consequent increase in damaging ultraviolet radiation at the ground. Although tentative, this suggestion triggered a political firestorm and contributed to the decision by Congress not to fund the supersonic transport. At about the same time Paul Crutzen and Harold Johnston showed that nitrogen oxides can act catalytically to destroy ozone in the stratosphere. Funding for research in stratospheric ozone increased dramatically, and in 1974 Mario Molina and Sherwood Rowland discovered that chlorofluorocarbon gases (CFCs) released at the ground

---

[49] Dave Atlas deserves special recognition for his unique mixture of ability and energy, integrity, and devotion to scientific ideals, demonstrated especially in his leadership of the NCAR National Hail Research Experiment. He enjoys arguing, but it is clear-headed probing for deeper questions and answers that drives him. I recall this quality from his days as a student at NYU.

can destroy ozone in the stratosphere. Molina, Rowland, and Crutzen, formerly an NCAR scientist, were awarded the 1995 Noble Prize for Chemistry for their work on stratospheric ozone.

The NAS carried out reviews of the subject in 1975, 1976, 1979, and 1982 that were independent of NAS/CAS.[50] Following the 1976 review government agencies acted with unusual speed to ban use of CFCs in spray cans, and legislation was adopted in 1977 authorizing the Environmental Protection Agency (EPA) to regulate any substance anticipated to affect stratospheric ozone. These actions were taken simply on the basis of scientific calculations; there was no observational evidence of ozone depletion.

At a NAS/CAS meeting in March 1963 committee member Edward Teller proposed support for a NATO study of weather control that Teller had earlier proposed to the NATO Science Committee. NAS/CAS members rejected Teller's proposal because it would compromise east–west cooperation and information exchange and because the scientific base for such a study did not exist. Later, during the Vietnam War, Teller proposed use of cloud seeding for military purposes in Vietnam. He was responding to a claim made by a group at the China Lake, California, Naval Laboratory that the Ho Chi Minh Trail could be made unusable through cloud seeding. I opposed the proposal, as did several others, because it would jeopardize international scientific communication and exchange of environmental data. I also pointed out that the objectives were far beyond proven capabilities. Jule Charney then spoke forcefully in opposition, and Teller said he knew when he could not win and would withdraw his proposal. Although Teller's views on matters bearing on politics were usually far from those of most of the committee, he was always pleasant and friendly at the few meetings he attended.

Throughout most of the 1960s NAS/CAS was chaired by Tom Malone who recognized with unsurpassed perception the role that the Academy could play in steering progress in science and technology, and in institutional development. He guided the committee brilliantly in

---

[50] Hans Panofsky, a member of the panel that produced the 1976 report, inadvertently revealed to a reporter a conclusion before the report was issued, provoking a torrent of poorly informed publicity. Panofsky offered to resign, but the offer was rejected, and no real harm was done. The publicity may have helped in attracting attention to the report when it was issued.

each of these areas and was responsible for many critical policy decisions that were visible only to those members of the scientific and government communities who were involved in these decisions. Some of his contributions to evolution of the atmospheric sciences are cited in other chapters of this memoir, but many contributions are not cited.

As the end of the decade approached, Malone felt it was time to move on to other activities and asked me to take over as NAS/CAS chairman, while agreeing to stay on as a member. I was appointed to the post by President Fred Seitz in January 1969. This coincided with reorganization of Academy committees in which NAS/CAS and other disciplinary committees were to operate within the National Research Council (NRC), and a system of rotation of membership was instituted. I had the task of introducing a three-year rotation schedule and informing several who had been members from the beginning of the committee that they were scheduled for rotation. Edward Teller was one of these.

In discussing future committee activities I recalled that at the beginning of the 1960s the Petterssen report had set the stage for the next decade, for expansion of research opportunities, and especially for the steps that led to the World Weather Watch and the Global Atmospheric Research Programme. As we approached the end of the decade and looked forward to the next, it seemed the proper time for a broad assessment of the current state of the atmospheric sciences and for identifying major objectives for the 1970s.

Highly rated proposals for research support and for important applications exceeded funds available in agency budgets, and the gap was projected to increase. At the same time societal issues had risen higher on the national agenda, while scientific research was becoming more difficult to justify to Congress. So agencies wanted guidance in setting research priorities. NAS/CAS agreed to my proposal to undertake a broad study addressing applications of the atmospheric sciences that could be expected to contribute most effectively to human needs in the decade, to suggest the outlines of realistic programs, and to recommend actions in priority order. Aeronomy and planetary atmospheres were not included because they had been subjects of recent studies by other NAS–NRC committees. Support was provided by eight federal agencies. Responsibility for organizing the study and for recommending actions was vested in a Steering Committee, and 11 subject-matter panels were appointed and charged with reviewing recent achievements and research plans and discussing potential applications to specific human needs.[51] John Sievers, as executive secretary, and Glenn Hilst, as study director, were responsible for much of the management and detailed preparation of the study. Individuals were assigned responsibility for reviewing national investments in scientific and technical resources relating to facilities, manpower, and research in government laboratories, universities, NCAR, and other institutions. Reports of the panels were prepared and distributed as preparation for a two-week summer study held at the Friday Harbor Laboratories of the University of Washington. More than 55 persons participated actively in the study; in addition representatives of the eight federal agencies supporting the study were invited and attended. Participants discussed panel reports, proposed priorities, and drafted recommendations.

The Steering Committee decided to use the voluminous panel reports as background and supporting materials, and to publish only a single, concise report summarizing the current state of the science, presenting recommendations for the 1970s in priority order, and estimating the needed resources. The 88-page report identified the following as deserving priority attention: short-range weather forecasts, dynamics of climate, and atmospheric chemistry, as well as continuing support for GARP and other initiatives from the 1960s (CAS/NRC 1971). New funding required by the recommended programs was estimated as about $450 million during the decade; this can be compared to the 1970 research budgets for meteorology and meteorological satellites that amounted to about $130 million.

I wanted to encourage development of a stronger role for atmospheric chemistry within the atmospheric sciences and for linking the work of chemists and meteorologists more effectively. I talked with Richard Craig about his serving as chairman of a panel to address these problems. I had known Dick Craig for twenty years or so and had come to admire him as exceptionally intelligent, well informed, and judicious, the qualities needed to lead the study. We believed that a useful outcome required that the panel include chemists and meteorologists with broad interdisciplinary perspectives. Dick agreed to serve as chairman, and we selected panel members who we thought could talk to each other, rather than past each other, as had occurred on several

---

[51] The Steering Committee consisted of Louis Battan, George Benton, Gordon Little, Tom Malone, and me as chairman.

earlier occasions when chemists and meteorologists had met to address interdisciplinary problems. The panel called attention to the fact that the nation was not prepared to respond to crises requiring understanding of the chemistry of the atmosphere and to the need for interdisciplinary training. The report identified the following as deserving priority attention: urban air quality, stratospheric composition, migration and fate of trace substances, and chemical aspects of climate change. Although the work of the panel was completed and its report approved by NAS/CAS in 1973, publication of the report was delayed until 1975, just as excitement over CFCs was reaching a crescendo (CAS/NRC 1975). The Academy conducted other studies limited to effects of CFCs that received intense scrutiny and interest, and the delay in release of the NAS/CAS report probably resulted from the attention these other reports received within the Academy review structure. A consequence of the delay was that the NAS/CAS report received less attention than it deserved.

In 1973 I asked Louis Battan to replace me as chairman of NAS/CAS but agreed to stay on as a member for a three-year term. Under Lou's leadership the committee directed attention to filling the gap between scientific reports and the popular press. A report was prepared that focused on the relation of the atmospheric sciences to society and was intended to have a broader readership than earlier committee reports (CAS/NRC 1977). I wrote the final chapter on Planning and Managing Atmospheric Research. That chapter encountered criticism from several agency administrators who evidently were offended by what they regarded as encroachments on agency territory. This response, although it may be understandable, seems not to recognize that Academy reports provide external perspective very important to interagency programs and in many cases strengthens agency credibility before Congress.

In 1976 George Benton took over as chairman, but had to resign a year later when he became associate administrator of NOAA. Charles Leith became chairman in 1977; under his leadership in 1978 the committee carried out an Atmospheric Research Review at Snowmass, Colorado, addressing priority research for the 1980s. I served on the steering committee for the review and participated in the Snowmass session as leader of the Task Force on Impacts of Weather and Climate on Society. Dick Reed succeeded Leith as chairman of NAS/CAS and presided over its incorporation within the Board on the Atmospheric Sciences and Climate when this broader structure was created in the early 1980s.

Other Academy studies in which I participated provided broad but limited contact with other disciplines and other communities of scientists. Included was the Physics Survey of the early 1970s that included a chapter on earth and planetary physics (Physics Survey Committee/NRC 1973), review of plans for the Lower Atmosphere Research Satellite System, planning for the International Decade of Ocean Exploration, coastal science—policy interactions, offshore oil exploration off the Washington and Oregon coasts, earthquake engineering, priorities in space research, and service on the Executive Committee of the Assembly of Mathematics and Physical Sciences, described later in chapter 13. So I had the opportunities to become familiar with a broad range of geophysical fields and programs.

CHAPTER 11

UCAR–NCAR Responsibilities

My associations with UCAR and NCAR extended over more than 30 years, from the 1958 meeting of university representatives that established the University Committee on Atmospheric Research, described in chapter 4, to the UCAR Committee on Membership that I chaired from 1986 to 1989. While at OST in the early 1960s, described in chapter 5, I had been an outside observer, critic, and defender of NCAR. From 1964 to the late 1980s I was involved in governance of UCAR. Over the years seven of my former graduate students have served on the NCAR research staff.

The UCAR constitution originally specified that each member university of UCAR be represented on the Board of Trustees by two member representatives, one a scientist and the other an administrator. By 1964 a growing number of the scientist member representatives felt that the Board of Trustees was heavily weighted toward administration and that it was poorly informed and insensitive on science matters. There were concerns within the NSF Board, OST, and in the university community that NCAR may have been growing too rapidly and with too little attention to the central objectives of the institution. In response to these concerns, in October 1964 the board created the Council of Members, consisting of one scientist from each member university, to better express the views of scientists to the board. I was appointed to represent the UW. The board chairman was specified also as council chairman. Unfortunately, this action emphasized the subordinate position of science in UCAR structure, thus tending to frustrate what the council had been created to achieve.

The council was delegated responsibility for annual review of NCAR's performance. I was appointed chairman of the first Evaluation and Goals (E and G) Committee, and I gained approval for a committee membership that could include scientists from non-UCAR universities. Louis Battan, Richard Goody, Colin Hines, Norman Phillips, Owen Phillips, and Harold Zirin were appointed to the committee.[52]

---

[52] Goody and Zirin were from Harvard and California Institute of Technology, respectively; at that time neither university was a member of UCAR.

The E and G Committee set to work in establishing procedures for general review of NCAR progress and for specific review of one-third of the NCAR program each year, and began the initial review. On April 14, 1965, the Board of Trustees, responding to political currents flowing in Washington, authorized the director to appoint an associate director responsible for "the area of application of the knowledge of the atmospheric sciences." The E and G Committee was dismayed by this action and promptly reported to the council its recommendation for delay until "the council shall be able to determine desirable objectives of NCAR in its next stage of development." The committee's first annual report, submitted in March 1966, emphasized that NCAR's central focus must be on carrying out first-class research on central atmospheric problems, and it recommended against forming "a separate administrative unit devoted to applications."

Clearly, the council and board had sharply different views of NCAR's function and structure. The E and G Committee report criticized the aerosol and trace gases programs rather severely, praised the solar physics program but called attention to lack of progress in developing an aeronomy program. The most important general conclusions were 1) that NCAR had been "slow to develop strong research programs directed at problems requiring personnel or facilities beyond the capacity of individual universities," and 2) a rank structure was needed together with procedures to achieve regular turnover of the scientific staff. The committee debated internally how strongly worded and unequivocal the report should be. Goody especially felt that we should leave no room for alternate interpretation or delay in implementing corrective actions, whereas I and most of the rest of the committee felt that this could be destructive and that it was better to state our conclusions and recommendations clearly but to give management freedom to improve things gradually. I was especially sensitive to Walt Roberts strong support by the board, which included the highly respected and influential, William T. Golden, and within political circles and the effect that highly visible turmoil within UCAR would have in Washington. In the end Goody probably felt that I had missed a valuable opportunity. He may have been right.

In 1966 the board granted the council the right to elect its own chairman; and I was elected to that position, while Richard Kassander, chairman of the board, was elected as vice chairman. The board appointed Walt Roberts as president of UCAR and John Firor as director of NCAR, separating corporate functions from direction of the research program. Norman Phillips succeeded me as chairman of the Evaluation and Goals Committee; in 1967 the committee reported that the program of dynamic and synoptic aspects of atmospheric circulation was not up to the high standards expected of NCAR, and it reiterated the need for developing a program of aeronomy.[53] The committee again emphasized the importance of a strong research program and the need for turnover of scientific staff.

The council also adopted resolutions that intruded on board territory, or that were so interpreted. The board responded by reducing the size of the board, introducing election of board members to terms of three years, and abolishing the council. An agreement was forged whereby NSF and UCAR were to be jointly responsible for appointment of review committees. Over the following several years it seemed necessary each year to adopt a new mode for conduct of the review function, and there was a troubled sense in NSF and in parts of the university community that more far-reaching changes were needed. During this time and in later years John Firor's keen insight and his ability to articulate issues of both science and policy were vital in maintaining communication between NCAR and NSF and between NCAR and UCAR member representatives.

In 1970 I was elected to the Board of Trustees; because I had been a visible irritant to some of the member representatives and the board, I felt that this indicated that need for further change was gradually being accepted. Walt Roberts proposed that he appoint a UCAR vice president and announced that he would resign when he reached 60 in 1975. This step did not represent a clear break from the past, and it was not implemented. My files for these years reveal a continuing exchange of letters between Walt Roberts and me on many proposed changes.

When Tom Malone was elected chairman of the board in 1972, he promptly set in motion steps toward further reorganization of UCAR's structure and procedures. Roberts offered his resignation effective on appointment of his replacement. Malone appointed a search committee with himself as chairman, and I served as a member of the committee. Names were proposed by committee members that indicated a wide range of views of what was needed in the presidency. I and others be-

---

[53] As noted in chapter 7, the initial attempt to develop aeronomy as an effective part of Atmospheric Sciences at the UW also failed—at just about this same time.

lieved that the situation called for a single leader with strong science credentials and that this could be achieved best by combining the positions of president and director. I proposed Francis Bretherton of Johns Hopkins University as UCAR president and as NCAR director. Bretherton was highly qualified by research achievement and leadership, but had little administrative experience. The committee ultimately agreed to recommend his appointment, and he was appointed to the two positions in October 1973. For the first year he divided his time between UCAR–NCAR and Johns Hopkins, where he still had commitments.

At the same time I was reelected to the board and was appointed chairman of the UCAR Organization Committee, a blue ribbon committee representing a broad range of constituencies and consisting of Jean Allard, David Atlas, Ray Chamberlain, Robert Charpie, William E. Gordon, Wytze Gorter, Henry Houghton, Clifford Murino, Robert M. White, with ex officio members Francis Bretherton and John L.J. Hart (legal counsel). Over a period of about eight months beginning in September 1973 the committee worked its way through several draft reports expressing a variety of positions on controversial issues. Unanimous agreement was finally reached on a report recommending changes to create a stronger, more centralized management structure. Recommendations included appointing an executive director of NCAR and a special assistant to the UCAR president for university relations, appointing a science advisory committee to advise the president and director, strengthening the board's executive committee and reducing the number of board committees from nine to five, and opening possible nomination for the board to faculty of universities beyond those who were member representatives, a provision I urged strongly though I realized it would meet substantial opposition. In voting on the committee's recommendations the member representatives specified that a nominee for the board who was not a member representative must be approved by the president of the nominee's university. With that single change, the recommendations were approved. This action could be viewed as completing the task begun some eight years earlier by the first E and G Committee. Early in 1974 Francis Bretherton appointed John Firor as executive director of NCAR, providing institutional memory and insuring effective broad communication.

Bretherton immediately provided a visible (and audible) emphasis on quality research as NCAR's central objective. He initiated a program of physical oceanography, substantially enhancing NCAR's potential strength in global aspects of atmospheric circulation. He was a hands-on director and involved himself directly in research programs and was forthright and perhaps sometimes insensitive in his assessments. His enthusiasm for participating in research may have made it difficult at times for him to separate his research interests from his responsibilities as NCAR director.

In January 1975 I became chairman of the board and immediately encountered an especially difficult and stressful issue. Political pressures had led earlier to establishment of a new NSF division called Research Applied to National Needs (RANN), and in 1972 NCAR had undertaken a 5-year RANN project [the National Hail Research Experiment (NHRE)] with the objective of determining whether hail storm damage could be significantly reduced by cloud seeding. By 1975 interim field results provided no evidence that hail had been reduced and indicated that a definitive result would require a longer period and more resources than were available. Director David Atlas recommended that the project objective be focused toward basic research on hail storms. RANN was unwilling to allow this and proposed management changes unacceptable to Dave. The UCAR board convened a panel of distinguished cloud physicists to provide guidance. Tom Malone and I, representing the board, attended as ex officio members. The panel concluded that, as the most credible and most visible cloud seeding project in the world, NHRE could not be terminated at that point; and Tom and I agreed. Each of the panel members had great respect for Dave's stand in this matter, and the conclusion was reached with great distress. We had to recognize that politics had won over science, not for the first time. Dave resigned as director and was replaced by Donald Veal of the University of Wyoming. In 1976 Dave was appointed as the first director of NASA's Goddard Laboratory of Atmospheric Sciences. I took some silent satisfaction in noting that my efforts of 1964–66 to encourage NASA to develop a strong atmospheric research program had finally been realized, for reasons independent of my earlier efforts.

I served as board chair for two years and then for a year as chair of the Budget Committee. For the most part I recall those years as stimulating and productive. I was often cast in the role of adjudicator or interpreter. The board had a clear focus, and issues were discussed, always directly and sometimes intensely, reflecting especially Bretherton's insights and initiative. The annual review of NCAR research programs continued to arouse criticism and controversy and sometimes exposed conflicts between university and NCAR scientists. Improved procedures for internal review and promotion were established. The rank structure for the staff was developed more fully, and a new category of senior specialist was created. A contract was negotiated for purchase of the 5th generation Cray-1 computer and the computer installed. Loss of some of the leading NCAR scientists and staff morale generally were matters for concern, to which the board tried to be responsive. We conducted a discussion including Ed Danielsen, Douglas Lilly, and perhaps others who had left NCAR; after that I tended to believe the dissatisfactions were an inevitable consequence of the changed institutional focus that could be overcome only by recruitment. As I recall, communications with NSF were chronically difficult, perhaps because Francis Bretherton's free swinging style did not fit comfortably in the bureaucratic mold, or because NSF staff found it difficult to monitor the extensive and increasingly sophisticated NCAR program. I thought the latter was the chief reason.

Loss of scientists who preferred more independence contributed to tensions within the staff that led Francis Bretherton to resign in 1979. He stayed on as an active senior scientist, and a few years later accepted appointment as director of the Space Science and Engineering Center at the University of Wisconsin.

My overall assessment of the roles of Walt Roberts and Francis Bretherton in development of UCAR and NCAR is that each contributed vitally and uniquely. Roberts's selection as the first director of NCAR placed first importance on mustering the support from the diverse sources needed to establish the institution, and it also reflected the thin ranks of the atmospheric sciences at the time. Walt's wide contacts with influential figures in government, business, and law were crucial in opening doors initially, in giving NCAR wide visibility, and in securing the site for the NCAR building; and his enthusiasm and generous spirit made the early UCAR–NCAR a family. He was chosen as the first director even though his background did not provide extensive familiarity with the atmospheric sciences. I believe that Walt Roberts's contribution was profound and essential to the institution's future. One must recognize at the same time that the NCAR building, designed around isolated towers, reflects Walt's vision for the institution as a collection of researchers pursuing individual projects of their own choice. That constitutes a continuing cost that should be weighed against the inspiring beauty of the building and its site. The dissatisfactions that developed around the program were inevitable; although they lasted too long, they were not fatal.

Francis Bretherton brought a new clarity of vision to NCAR and a restless urgency to get on with research. Francis inevitably disturbed some of the staff who were accustomed to Roberts's style, and his method of personal communication through challenging questions, stimulating to some, was upsetting to others. NCAR matured during his tenure; and many scientific careers, including some prestigious ones, were enhanced by several years spent there. I believe that Francis was responsible for changing NCAR's focus and strengthening the research foundation of the institution, critical steps in securing national and international recognition of UCAR–NCAR's unique role.

After leaving the board in 1978 I continued a fairly active involvement in UCAR matters: review, nominating, selection, and membership committees of various sorts. In 1979, following Bretherton's resignation, there was interest in UCAR moving beyond management of NCAR to undertake a larger role in organizing and supporting multi-institution projects, thus introducing major changes in institutional objectives and structure. To examine this proposition the Members Advisory Committee on UCAR Initiatives was appointed, and I was asked to serve as chair. We cautiously supported the concept of extending UCAR's role. When Bob White became president of UCAR in 1980 he moved quickly and vigorously to extend UCAR's responsibilities well beyond managing NCAR. Wilmot Hess was appointed as NCAR director, separating clearly the responsibilities of the UCAR president and the NCAR director. Under UCAR a number of Cooperative University Programs (CUPS) were introduced—they included the Office of Interdisciplinary Earth Studies, the Corporate Affiliates Program, the International Affiliates Program, and the Cooperative Program for Operational Meteorology, Education, and Training (COMET).[54] More recently, reorganization has created the UCAR Office of Programs (UOP); within UOP are COMET and seven other programs that include the activities identified above.

In 1983 I was nominated for another term on the board but was not elected. A new generation of member representatives evidently felt it was time for new blood on the board, but I wondered also whether the vote represented disapproval of policies I had espoused, especially

---

[54] Within NCAR the Environmental and Societal Impacts Group had been created in the 1960s. With Michael Glantz as director, it has made major contributions to application of research results to understanding impacts of climate change.

opening board eligibility beyond member representatives, which had been opposed fairly widely. Maybe both factors were present.

From its beginning UCAR had avoided a visible role in science policy except as NCAR operations were directly affected by policy. UCAR responded to requests for information, but was careful not to express views that might run counter to those of NSF or of its member universities. During Bob White's two years as UCAR president an office was opened in Washington, D.C., to facilitate contacts with agencies and the Congress; and Skip Spensely, an experienced and astute Congressional staffer, was added to the UCAR staff. Beginning in 1984 George Benton led the new Committee on Public Education and Public Policy (COPEPP), of which I was a member, in exploring how a coherent consensus on important policy matters might be conveyed to appropriate officials and committees.[55] COPEPP carried out several small projects of this sort; but the difficulties were great and, although there were no explicit failures, the effort was abandoned after three or four years. The lesson learned here was that UCAR could facilitate communication and increase its understanding of policy matters through staff employed for those purposes, but that to advocate policy on behalf of the corporation and its members was very difficult and potentially dangerous. That conclusion tends to keep policy issues close to the UCAR president's office, and that seems to be the route that Richard Anthes has wisely followed as UCAR president.

By the 1980s the original UCAR membership of 14 universities had more than quadrupled, and the membership included a far broader spectrum of educational and research programs. Criteria for membership had evolved over time, and in some cases university programs had changed so that questions of membership qualifications arose. From 1986 to 1989 I served as chair of the Committee on Membership; during that time we tried to establish criteria that were both flexible and objective; and in some cases we tried by direct contacts with university administrations to encourage development of stronger programs. I believe we established much needed standards for membership and procedures for periodic review and criteria for continuing membership. But to maintain standards of membership requires frequent and consistent attention—the job is never done.

---

[55] Deborah Stirling, experienced as a Congressional staffer, served on the UCAR staff and was a major contributor to COPEPP's work.

CHAPTER 12

AMS Activities

As noted earlier, I became an associate editor of the *Journal of Meteorology* in 1950. That began an association with the American Meteorological Society (AMS) that extended more or less continuously into the 1990s. I was elected to a three-year term as councilor in 1957 (noted in chapter 4) and was elected to another council term in 1973. From 1965 to 1969 I served as commissioner of the Scientific and Technological Activities Commission (STAC). This was a busy period during which STAC committees proliferated, and many committees held increasing numbers of symposia at meetings of the AMS and were active in other ways. I was concerned that some of the committees had become insular and parochial; and I wanted to stimulate and strengthen STAC by encouraging turnover and enlisting active, younger members of the research and professional communities; emphasizing the need for more uniform standards; reducing the number of committees; and promoting linkages among STAC committees. In 1966 I proposed reorganization of STAC, but the council was reluctant to stir the pot as vigorously as my proposal may have implied. After extended discussion agreement was finally reached by the council on a request to the STAC Commissioner to "prepare a desirable plan for reorganization." My intention had been only to keep the council informed of what I regarded as evolutionary changes, not to open a continuing negotiation with the council. Rather than follow that course, I simply asked the various committees to prepare frames of reference. This was accomplished, providing a view into the structure and operation of each committee and giving each an opportunity for self-appraisal. The result was gradual movement toward my original goal; the course followed was probably better than the one I had originally proposed.

Historically, the Society had avoided taking positions on policy issues or had dealt with such issues hesitantly and reluctantly, reflecting the fact that many members were employed by government agencies. As interest in environmental problems grew during the 1960s policy issues came before the AMS with increased force. Establishment of the Environmental Science Services Administration in 1965 enhanced visibility of environmental concerns and helped to validate them within the general membership of the AMS. As an explicit example, use of weather modification as a weapon in the Vietnam War was discussed within the military and by the media and the general public and ultimately came before the AMS Council.

In 1969, on the recommendation of Planning Commissioner Robert M. White, the Committee on Public Policy (COMPUP) was created to propose statements expressing AMS positions on policy issues. Statements concerned with maintenance and improvement of man's environment and other issues were proposed, but the council remained reluctant to step into the arena of formal statements on national policy.[56]

In 1972 the AMS Executive Committee adopted a statement urging "the United States Government to present for adoption by the United Nations General Assembly a resolution pledging all nations to refrain from using weather modification for hostile purposes." And following submission of Senate Resolution 281 by Senator Claiborne Pell and 13 others, opposing "use of any environmental or geophysical modification activity as a weapon of war," hearings were held under the Senate Foreign Relations Committee at which Dick Reed, as AMS president, and other meteorologists testified in support of the resolution. President Reed read the statement adopted by the Executive Committee.[57] However, the statement does not appear in minutes of the council published in the *Bulletin*, indicating that it may not have been approved by the council.

Also in 1972 AMS President Dick Reed, with council approval, took the opportunity of President Nixon's announced visit to China to request the president to convey to representatives of the Chinese meteorological community the Society's desire to establish closer relationships between American and Chinese meteorologists. In pursuit of this

---

[56] NAS/CAS was much less hesitant to address controversial topics. For example, the report of the 1970 Friday Harbor study (NAS/CAS 1971) recommended that the U.S. government present to the United Nations a resolution "dedicating all weather-modification efforts to peaceful purposes and establishing . . . an advisory mechanism for consideration of weather modification problems of potential international concern before they reach critical levels."

[57] Testimony appears in U.S. Senate Committee on Foreign Relations United States Senate, Subcommittee on Oceans and International Environment (1972). Testimony included a letter from National Academy of Sciences President Phillip Handler citing extensively from the 1971 NAS/CAS report on *The Atmospheric Sciences and Man's Needs* in support for the Senate resolution.

objective, the Society offered to send a delegation to China. The reply to Dick Reed's letter encouraged the AMS to pursue the project on a nongovernmental basis. Two years were required to bring the visit about, but it finally occurred in April and May, 1974 and led to succeeding visits by Chinese delegations and further visits to China by U.S. individuals and groups (Kellogg et al. 1974). Success of this initiative helped to open the door toward fuller participation of the AMS in the public policy arena.

During the decade of the 1970s atmospheric policy issues became more prominent and more numerous, the result of rapid growth of the science and the profession and increased relevance of the field to national objectives and needs. Reflecting these changes, policy symposia were held at AMS national meetings in 1976 and 1978. I participated in each and in 1978 initiated and, as noted in chapter 9, secured funding for a third symposium to which 13 invited individuals addressed critical topics. The central conclusion reached by the participants was that the prospect of continued growth of industrialization and world population require closer collaboration of scientists and policy makers and development of new institutions that can respond to both the results of scientific research and national objectives.

In 1979 I was elected to the presidency of AMS for the year 1981 and therefore became a member of the Executive Committee for the period from 1980 to 1984. A lawsuit had been brought in the late 1970s challenging the Society's procedures for approving applications as Certified Consulting Meteorologists (CCMs). Legal negotiations over several years were distracting and unpleasant, but resolution of the suit early in 1981 by a consent decree resulted in procedures of greater regularity and objectivity and created a stronger CCM program. Also in 1981 the Society submitted an amicus curiae brief in an appeal to a decision that had found the American Association of Mechanical Engineers legally responsible for a statement made by a member who was serving as a member of a committee of the association. If upheld, this decision could have made it virtually impossible for AMS committees to operate. Fortunately, the decision was overturned.

At the beginning of my presidency in January 1981 I outlined five objectives that the Society needed to address in the coming year: 1) strengthening links with the physical oceanography community as it sought to create its own society, 2) moving toward publication of a *Journal of Climate*, 3) strengthening the interface between private meteorology and government in response to technological developments in observation and data processing, 4) seeking ways to expand opportunities for minorities and women, 5) preparing to respond to the competition between the national objectives of environmental protection and resource development that was likely to result from changed policies introduced by the new Reagan administration.

In 1981 the Society began publication of the *Journal of Climate and Applied Meteorology*, the origin of the present *Journal of Climate*.[58] The *Journal of Physical Oceanography* had become a monthly publication in 1980, and in 1981 I appointed the Committee on Oceanography with Jim Baker as chair to propose further steps relating to the role of physical oceanography within the Society.

Especially critical policy issues affecting the AMS and the meteorological community generally arose at the beginning of the Reagan administration in 1981. I appointed a Committee on Impacts of Budget Changes with Charles Hosler as chair to advise the Executive Committee on effects of federal budget actions on AMS programs and concerns and a Committee on Private Sector Activities of the AMS with George Benton as chair.

Attempts were made by the administration to replace some of NOAA's senior professional leaders with political supporters. It appeared that Dick Hallgren, director of the National Weather Service and others were being targeted. As president of the AMS I urged John Byrne, NOAA administrator, to resist these moves. In many contacts Byrne expressed his determination to defend scientific standards; at the same time he was severely limited in his freedom of action. In 1982 at Bob White's initiative a letter from former AMS presidents was sent to key officials emphasizing the importance of retaining NOAA leaders of demonstrated qualifications and performance. And prominent Congressional leaders opposed the administration's ambitions. Together, these actions were successful; NOAA leaders retained their positions, but the threat took its toll on agency morale.

Proposed actions of the Reagan administration, described in chapter 9, included sale of the weather satellites to private industry, reduction of Weather Service functions to statutory requirements, assignment of research aircraft to NOAA headquarters, and centralization of Weather Service computers under the Department of Commerce. In

---

[58] The *Journal of Applied Meteorology* had originated earlier, and its former name was restored when the *Journal of Climate* was spun off in 1988.

1983 the AMS Council responded by adopting a statement describing effects that sale of the satellites would have on U.S. access to international meteorological data and expressing the Society's concern regarding the proposed sale of satellites to private industry. I was invited to testify at hearings on authorization of the NOAA budget before the Senate Committee on Science, Technology, and Transportation and, as noted in chapter 9, took that opportunity to criticize in detail administration plans for drastic reduction in NOAA's capabilities (U.S. Senate Committee on Commerce, Science, and Transportation 1983). And testimony on the importance to the nation of the Weather Service by George Benton and Conley Ward at hearings of the House Committee on Science and Technology was published in the AMS *Bulletin*.

In the early 1980s environmental effects of nuclear war became a focus of heightened concern by the science community and the general public. Papers published in 1982 and 1983 emphasized that a nuclear war between the United States and USSR could result in disastrous global environmental consequences, and the concept of "nuclear winter" attracted extraordinary attention (Crutzen and Birks 1982; Turco et al. 1983). Resolutions calling for renewed efforts to reduce the threat of nuclear war were adopted by ICSU (September 23, 1981), by the Council of the American Association for the Advancement of Science (AAAS) (January 7, 1982), by the NAS (April 27, 1982), and by the American Physical Society (January 23, 1983). A symposium on environmental effects of nuclear war, led by Julius London and Gilbert F. White, was held by the AAAS on May 28, 1983 and later published (London and White 1984). Following the AAAS symposium Thomas Malone, Julius London, and Chester Newton, proposed that the AMS Council issue a formal statement adding its voice to those emphasizing actions to prevent nuclear war. The Executive Committee approved this initiative; and the council, under strong urging by me and by other councilors and after debate expressing reservations of some members, adopted a statement referring to "effects propagated through the atmosphere to the entire globe that could cause the destruction of the biological base that sustains human life" and concluded by "calling on the nations of the world to take whatever steps are necessary . . . to prevent the use of nuclear weapons and avoid nuclear war." To my knowledge this was the strongest and most unequivocal response that had been taken by the council to a major policy issue.

Also in 1983 the Society convened a Forum on Meteorology and the Future with the unstated but clear purpose of presenting to the nation an alternative to the minimalist weather service advocated by the administration. The forum produced an Agenda for Action calling for the nation to make an unreserved commitment to a new weather and climate agenda. It declared that a partnership of effort among universities, government, and the private sector could bring tangible benefits to the national economy and enhance protection of life and property and contribute to preservation of environmental quality. The agenda recognized that wise policy decisions by the Congress and the Executive Branch were required. One cannot claim that the weather service envisioned by the forum was achieved, or even approached; but the forum may have helped the Congress in its effort to protect NOAA and its Weather Service from even worse ravages by the administration.

My term as president-elect, president, and past-president witnessed substantial progress in the ability of AMS to respond to important policy matters. It was a stimulating time, and many individuals and groups shared in responsibility for the positive actions taken.

In April 1982 I represented the AMS at the Centenary Meeting of the Japan Meteorology Society in Tokyo. Tu-Cheng Yeh and I gave speeches congratulating the Society on behalf of the Chinese Meteorological Society and the AMS. Yeh had gotten his Ph.D. at the University of Chicago in the late 1940s, and we had met at earlier international meetings. Other speeches were made by Professor K. Gambo, president of the Meteorology Society of Japan, Professor Tomio Asai of Tokyo University, and by Japanese government officials. In my remarks I expressed special appreciation of the AMS and the U.S. meteorological community for the crucial role played during the Global Weather Experiment by the Japanese geosynchronous satellite *Himawari*, "sunflower" in Japanese, that had provided observations of clouds over the western Pacific. There was a commemorative ceremony, and Professor Asai arranged an elegant dinner for Yeh and me as well as visits to the Meteorological Agency and the Meteorological Satellite Center. During the four days of our visit Yeh proved to be a highly skilled science diplomat and a friendly and stimulating companion.

In the early 1990s I served on the Committee on History of the Atmospheric Sciences and as chair from 1992 to 1995. A one-day symposium was presented at the 1992 annual AMS meeting on Historical Currents in Meteorology. Nine papers were presented on major advances in observation, theory, and professional practice. At the 75th Anniversary Meeting of the AMS in 1995 a one-day symposium focused on progress since 1920. Edward Lorenz spoke on advances in dynamic meteorology; Robert Serafin discussed techniques of observation; George Cressman, forecasting; Michael Prather (substituting for Ralph Cicerone), atmospheric chemistry; Mark de Maria, weather modification; John Kutzbach, climate; Norbert Untersteiner, Arctic meteorology; and John M. Wallace, the El Niño. These two symposia responded to and nourished growing interest in the historical roots of the atmospheric sciences.

CHAPTER 13

The National Climate Program

The objective of achieving "a better understanding of the processes that determine the system of world climate" was included in the 1961 NAS/CAS report and in the U.N. resolutions that led directly to the World Weather Watch and GARP. And two objectives were stated in the 1969 agreement between WMO and ICSU that launched GARP—the first, "increasing the accuracy of forecasting over periods from one day to several weeks"—and the second, "better understanding of the physical basis of climate." Although the first objective received nearly exclusive attention throughout most of the 1960s, by late in the decade the second objective began to gain attention, stimulated especially by Keeling's evidence that atmospheric carbon dioxide was increasing inexorably. As noted in chapter 10, the 1970 NAS/CAS summer study at Friday Harbor recommended that "increased efforts should be directed toward understanding the dynamics of climate" (NAS/CAS 1971). Later that same summer I participated in the Williams College summer study of long-term environmental problems, organized by Carroll L. Wilson of MIT. After the study the leaders concluded that climate change is probably the most threatening of the long-term environmental problems resulting from human actions [Study of Critical Environmental Problems (SCEP) 1970]. This led to a study of man's impact on climate in 1971 and to the 1972 U.N. Conference on the Human Environment, both held in Stockholm, Sweden, under international auspices. Drought in the Soviet Union and in the African Sahel, also in 1972, resulted in rise in grain prices and disrupted the world grain market; and climate change suddenly snapped into the line of sight of the U.S. Central Intelligence Agency and the Nixon White House.[59] Although the intense attention was short-lived, climate research began to receive increased support by NSF, NOAA, and other agencies. The U.N. Environmental Programme (UNEP) was established. In 1974 under

---

[59] Joseph O. Fletcher played the crucial role in gaining the attention of high-level officials for climate change.

the NAS Committee for GARP the Panel on Climate Variation issued a report that helped to establish climate change as a tractable and stimulating scientific endeavor and strongly influenced subsequent developments (U.S. Committee for GARP 1975).

In 1976 NAS President Phillip Handler appointed me to the Executive Committee of the Assembly of Mathematics and Physical Sciences (Ex-Com-AMPS), responsible for oversight of Academy committees and boards in those fields. Jack Oliver and Preston Cloud, both geologists, and I represented the geophysical sciences. Meetings were devoted to reviews, discussions, and approval of reports and new initiatives proposed by committees and boards. During my three-year term on Ex-Com-AMPS climate was the chief topic to which I contributed. I provided briefings and memoranda summarizing current understanding of climate issues. Discussions of the need for a NRC unit with a special focus on climate issues began during the summer of 1976, and at the meeting on November 25 I was asked to chair an ad hoc Subcommittee on Climate.[60] We developed a proposal for creation of a Climate Research Board (CRB), and the board was approved at the February 25 meeting, at just about the time that the White House announced that Bob White would be replaced by Richard Frank as administrator of NOAA, thus making White potentially available to serve as chairman of CRB. Discussions of this opportunity proceeded during the spring, and Bob White became the full-time "resident" chair of CRB on July 1, 1977. Funding for the CRB was provided by government agencies and the Sloan Foundation. White served as CRB chair for more than two years; during that time he played critical roles in formulating the U. S. national program, and in shaping the National Climate Act and the World Climate Programme.

During the 1977–78 year that I spent in Washington, D.C., as a consultant to NOAA, federal agencies were moving to stake out their claims to what would become the National Climate Program, and related bills were introduced in Congress. I had opportunities to follow these developments at close range and to comment on them to NOAA and the Climate Research Board. In March I was asked to testify at Congressional hearings on impacts of weather and climate changes on the economy, and took that opportunity to point out the need for more

---

[60] Joseph W. Chamberlain, Preston Cloud, Richard L. Garwin, and Allan R. Robinson were members of the subcommittee.

effective planning and coordination of the Climate Program (House Committee on the Budget 1978). And, as noted in chapter 9, my testimony was summarized in an editorial published in *Science* (Fleagle 1978).

In April 1978 the International Workshop on Climate Issues, a product of Bob White's first year as CRB chair, was held at the International Institute for Applied Systems Analysis in Laxenburg, Austria. The workshop reviewed scientific understanding, human influences on climate, and impacts of climate. Bob White chaired the workshop and invited me to attend as one of more than 300 scientists and government officials from more than 50 countries. The report set forth the need for and recommended a world climate program embracing research, data requirements, applications of climatic information, and impacts of climate change. The World Meteorological Organization responded to these recommendations a year later by launching the World Climate Programme (WCP) on behalf of its 149 member nations. The World Climate Research Programme (WCRP) became the centerpiece of the four components constituting the WCP.

A one-week workshop in which I participated was held by the Climate Research Board at Woods Hole, Massachusetts, in July 1978 that reviewed the research plans drawn up by federal agencies. The Department of Energy (DoE) had proposed a plan in which it was to be responsible for assessment of interagency research relating to greenhouse gases. The plan seemed ambitious for an agency with rather specialized and limited interest in the atmospheric sciences, and the workshop report described the DoE proposal as lacking in necessary linkage and coordination with other agencies. DoE subsequently instituted a program of research at universities and DoE laboratories that became highly effective in clarifying the role of carbon dioxide in the carbon cycle and in climate change. The Woods Hole workshop contributed significantly to the National Climate Act of 1978, and to the Climate Program developed under the act.

In 1980, as part of a Climate Research Board symposium, I undertook along with Allan Murphy a study of the use of climate information in management of the flow of the Columbia River. We concluded that as demands for hydropower, irrigation, fish survival, and other uses of the Columbia increase fluctuations in climate are likely to exacerbate problems of management, and that better use of climate information will become necessary. That study led to my recognizing that the most serious consequence of global warming in the Pacific Northwest is likely to be reduced winter snowpack in the mountains of the Columbia watershed. This became the chief subject of a paper published later (Fleagle 1991), as well as of a paper presented to the 16th Annual Symposium on Environmental Law at the University of Oregon (March 1998) and of op-ed pieces published in Seattle newspapers (Fleagle 1998).

Global Environmental Change

During the 1970s research had advanced the frontiers of understanding for a broad array of environmental problems: climate, stratospheric ozone, air quality, the carbon cycle, water resources, biological diversity, soil properties, and still others. For the most part each was studied independently of the others using more or less unique observation systems and by separate groups of scientists. However, it was evident that environmental problems were becoming more and more interdisciplinary and that observations were needed that could specify the whole phenomenon of interest, often on a sub-global or global scale, phenomena identified under "global environmental change."[61]

Two especially important interdisciplinary studies were carried out in the late 1970s under the ICSU Special Committee on Problems of the Environment (SCOPE) by committees of meteorologists and biologists led by Bert Bolin (Bolin et al. 1979, 1981). These studies advanced understanding of the role of the global carbon cycle in climate change and made clear the need for interdisciplinary research utilizing global observations.[62] In 1982 a workshop initiated by NASA and chaired by Richard Goody, outlined programs of space observation and research needed for study of an array of global environmental problems. A year later NASA's Earth Systems Science Committee, chaired by Francis Bretherton, carried out a comprehensive review of the sci-

---

[61] The U.S. has had a less important role in research programs addressing global environmental change than in the earlier research leading to recognition of this interdisciplinary focus. This reflects in part the global scale of these problems, but it also reflects changes in U.S. national policy instituted by the Reagan administration and the resultant failure of the U.S. to undertake a leadership role in addressing international environmental problems.

[62] The SCOPE studies of the carbon cycle can be traced to Carl Rossby's decision in the mid-1950s to develop at the University of Stockholm a graduate program focusing on global scale atmospheric chemistry. Bert Bolin was an early product of this unique program. He combined interest and understanding of the dynamics and the chemistry of the atmosphere; in later years, as Rossby's successor, he demonstrated unusual ability to inspire and lead his colleagues and provided essential leadership to a series of international programs.

ence of the earth as an integrated system linking the atmosphere, ocean, cryosphere, land surface, solid earth, and the biosphere. At the same time an Academy workshop chaired by Herbert Friedman of the Naval Research Laboratory (whom incidentally I had known since 1940 when he was completing his Ph.D. at Johns Hopkins) proposed an interdisciplinary program of a new order embracing the geophysical and biological sciences. These initiatives provided the conceptual basis that, together with the SCOPE reports on the carbon cycle, led in 1986 to creation within ICSU of the International Geosphere–Biosphere Programme: A Study of Global Change (IGBP) to undertake interdisciplinary research linking the geophysical and biological sciences. Core projects were launched in atmospheric chemistry, terrestrial ecosystems, the hydrologic cycle, ocean fluxes, and records of past climates. An extensive system of scientific and administrative committees was established bringing together biologists, geochemists, oceanographers, and meteorologists working on such problems as the role of dimethyl-sulfide produced in the ocean on cloud formation, the evolution and properties of atmospheric aerosol, the chemistry of ozone in the stratosphere and troposphere, and transfer of carbon dioxide between the atmosphere and ocean.

One might have expected the IGBP and the WCRP to be organized within a single international program, but this did not occur. The WCRP had been initiated and developed by physical scientists and it consequently focused on physical aspects of climate, while the IGBP responded to the interests of biologists and chemists in an array of environmental problems. The headquarters for the WCRP remained with the WMO in Geneva, Switzerland, while the headquarters for the IGBP was established in Stockholm, Sweden. Numerical climate modeling remained the centerpiece of the WCRP, while studies of biogeochemical cycles became the centerpiece of the IGBP. Communication channels were established, but research concerned with linkage of biogeochemical cycles with climate modeling was not a central focus for either.

On the policy front the U.N. Environmental Programme and specifically its executive director, Mostafa K.Tolba, took the lead in the 1980s in responding to the stratospheric ozone problem. Over the period from 1983 to 1987 the Montreal Protocol was drafted and received widespread international approval. As amended in 1989 it called for phase-out of consumption of chlorofluorocarbon gases (CFCs) by the year 2000. A consequence has been that, although stratospheric ozone continues to decrease globally, decrease is occurring at a decreasing rate;

and many observers are optimistic that it may be possible to contain the threat of increasing exposure of humans to dangerous UV radiation. The Montreal Protocol has been hailed as marking a new era of environmental statesmanship.[63]

Policy aspects of global warming are far more complex and difficult to resolve than those of stratospheric ozone. Actions to limit greenhouse gas emissions are opposed by many corporations involved in energy generation, distribution, and consumption, very powerful players on the political stage. And the impacts of global warming, as well as possible actions to mitigate them, are likely to fall very unequally on different nations and regions, making international agreements difficult to achieve.

In 1985 an international study of climate change under WMO, ICSU, and UNEP led to creation of the WMO Advisory Group on Greenhouse Gases with responsibility for assessing the science and for exploring the potential for a global convention on greenhouse gases and chaired by Bert Bolin. Growing interest in limiting emissions of greenhouse gases, amplified in the United States by the hot summer of 1988, caused concern at policy levels in the United States and other governments that the WMO Advisory Group could become a loose canon. To ensure governmental control, the United States and other governments moved to create under WMO and UNEP the Intergovernmental Panel on Climate Change (IPCC), with responsibility for scientific assessment and for reviewing impacts of climate change and methods of mitigation, but structured to insure government control of policy aspects. Bert Bolin was appointed as IPCC chairman. Beginning with its 1990 report IPCC scientific assessments have digested and summarized the many research studies embracing the various aspects of climate and through these reports have been responsible for disseminating the increasing understanding of climate change. At the same time, the IPCC has faced strong opposition on the policy front. The Global Climate Coalition, a powerful industry lobbying group, and other critics have deliberately confused scientific assessments and policy conclusions and have accused the IPCC of inconsistent statements. These efforts probably have been responsible for slowing adoption of effective emission limits. The international conferences at Rio de Janeiro, Brazil, in 1992; Kyoto,

---

[63] The story of the Montreal Protocol and its consequences has been told by Benedick (1991).

Japan, in 1997; Buenos Aires, Argentina, in 1998; and The Hague, Netherlands, in 2000 have made progress on international agreements but at a pace far slower than that projected by IPCC reports on policy and impacts. And the U.S. Congress and the George W. Bush administration have been unwilling to endorse even these agreements.

In the early 1980s it seemed increasingly clear to me that the way was opening to unprecedented achievements in interdisciplinary research and in application of results. At the same time I knew that standards for interdisciplinary research are not well defined, so that mediocre research can sometimes masquerade under the interdisciplinary banner. On the other hand, reliance on traditional reductionist methods sometimes can lead scientists to venture onto treacherous ground in interpreting their research results without thorough understanding of the interdisciplinary context.[64] To resolve these issues in my own mind I felt I needed to devote some concentrated study to atmospheric aspects of global environmental change. An opportunity arose in the early 1980s when I was asked by Glenn Hilst to serve as a member of an environmental advisory committee for the Electric Power Research Institute (EPRI). The committee's task was to help to develop an air quality research program. I successfully urged Morton Barad's appointment as a special consultant to carry out a survey of air quality activities at institutions around the world; he carried out the task with characteristic organized thoroughness and discernment . This provided a baseline from which to judge what might be appropriate and feasible at EPRI. I found the opportunities for discussing air quality research and policy with members of the committee who were experts in the subject, such as John Wyngaard and Ken Deirmenjian, stimulating; but the bulk of the work consisted of reviewing research proposals of narrow focus. Representing the committee, I served as a consultant to a national review of the National Acid Precipitation Assessment Program that gave me an opportunity to comment more directly and coherently on the need for changes in national air quality policy.[65] What effect my analysis had on policy remains a mystery.

---

[64] Some journals have chosen to publish articles of this kind, evidently believing that it is best to get new data and interpretations out for assessment and criticism. I am not convinced that this is efficient and prefer the policy of interdisciplinary review normally followed by *Science* and by AMS and AGU journals.

[65] The review was led by Robert Sievers of the University of Colorado.

Another opportunity for me to extend my understanding of global change arose through an exchange program between the UW and Evergreen College. I spent the spring quarter of 1985 at Evergreen in Olympia, Washington, teaching a course in atmospheric processes and global change. The students were specializing in environmental studies, and the course was at about the level appropriate to beginning juniors at the UW. The first half of the course included the physics, chemistry, and meteorology of climate, stratospheric ozone, and tropospheric air quality. The second half included the evidence for those aspects of global change and uncertainties associated with the evidence and with extrapolating past changes into the future. It gave me the chance to bring together what was known and what was unknown and to make an appraisal of institutional changes needed to understand global change more thoroughly.

Global change came into focus in still another way through President Reagan's 1983 Strategic Defense Initiative (SDI), creating a project intended to render intercontinental missiles "impotent and obsolete," an initiative opposed by the overwhelming majority of scientists and science administrators outside the Reagan administration and the defense industry. At the UW a group of faculty, organized by astronomer and physicist Paul Boynton, offered a course on the SDI. At separate sessions of this course J. Gregory Dash and Robert Williams from physics and Conway Leovy and I from atmospheric sciences analyzed the SDI from different perspectives. Speakers from off-campus included John Pike of the Federation of American Scientists and James Ionson, science director of SDI, and a Boeing employee. The last two gave a positive view of the president's proposal. Those of us opposed to SDI emphasized the environmental consequences of nuclear war and global threats to civilization. The course attracted a large class from faculty and students and from the off-campus public; response was animated and largely critical of SDI.

In the fall quarter of 1988 I organized an interdisciplinary seminar on climate change under the Institute for Environmental Studies and was able to invite leading scholars from other institutions to participate, as well as UW faculty. Fields represented included atmospheric sciences, biological sciences, economics, oceanography, and political science. The seminar attracted faculty and students from a wide range of departments and I believe was considered highly successful.

I retired a year later and began to direct special attention to the origins and background of the interagency U.S. Global Change Research

Program that had been launched in 1989. I was invited to participate in a discussion of Environmental Sciences and Public Policy at the 1991 AMS annual meeting and gave a paper on the U.S. Government Response to Global Change. I published several papers on this general subject in 1991 and 1992 (Fleagle 1991, 1992a,b), and Praeger Publishers proposed publication of a book on global environmental change. Their intentions sounded good, and I signed a contract to produce a manuscript by December 1993. I wish now that I had written the manuscript first, but at the time it was a relief not to have to wonder whether a publisher would be interested.

The book was addressed especially to government officials and staff concerned directly or indirectly with global change, but I hoped that it might also contribute to public understanding of these issues. The outline for the book fell into place quickly. The book would focus on institutional changes needed to respond to environmental problems of coming decades and would be concise, even terse. It would begin by relating the rise of public concern for environmental change to the changing place of science in society and in government. The evidence for global changes in climate, stratospheric ozone, and air quality would be reviewed and possibilities discussed for anticipating future changes. Some of the impacts of the changes would be discussed, and then the policy landscape would be described under three major components: academic, government, and nongovernmental organizations. Changes in the United States would be related to changes occurring internationally. The final chapter would be a discussion of institutional change required at the federal level to prepare for global environmental changes of the first decades of the twenty-first century. I hoped that it would fill what I saw as a gap in global change literature.

The book was published in the fall of 1994 (Fleagle 1994). Despite my effort to set a price that would make it attractive to a wide audience, it was priced at $59.95, severely limiting its distribution. The book was listed in the publisher's social science catalog, and was not made available for display at meetings of the AMS or the American Geophysical Union (AGU) in early 1995. A second printing was issued in 1996 at an even higher price. Neither printing has been marketed actively. Reviews have ranged widely from critical to laudatory. The most critical reviews have been written by political scientists, who, while finding some of the history useful, found the policy analysis inadequate. One criticized the attention given to evidence of global change and discussion of uncertainties because they are discussed extensively in IPCC reports (sic). A few reviews found the book broadly well informed, balanced, and notably valuable. And there were several unsolicited commendations from the UW Political Science Department.

More recently I helped to organize the Northwest Climate Change Council, a public interest group headed by lawyer Blair Henry and including academics, public officials, and business leaders. It has made effective contacts covering a wide range of stake holders, and has held a number of public forums. Establishment of the Washington Climate Center, a clearing house for climate change information and related activities, has been proposed to the state legislature and endorsed by vote of one house. I have spoken on issues of global change to public groups in Seattle and Port Townsend, Washington, and to the 16th Annual Environmental Law Conference at the University of Oregon. Work with a national student group, Ozone Action, may have helped to convince the Ford Motor Company to withdraw its membership from the industry lobbying group, the Global Climate Coalition. Work with a UW faculty–student group influenced the Board of Regents to begin discussion with corporations in which it has invested and that are major generators of greenhouse gases, looking toward changes in corporate policy or, alternatively, toward possible divestment of investments. And articles on the results of the Kyoto and Buenos Aires conferences have been published as op-ed pieces in the *Seattle Times* and the *Seattle Post Intelligencer*.

Other recent activities include service as one of many reviewers of the IPCC Third Assessment Report for Working Group I (Scientific Assessment) and Working Group III (Mitigation of Climate Change). This report was published in early 2001, five years after publication of the second report and ten years after the first report.

The small actions that I have been able to take toward better understanding of global change fall into line with similar actions by many of my colleagues and by others who have enlisted in the campaign. At the same time the media are doing an increasingly good job in explaining science and discussing science policy. The result has been a slowly maturing public understanding of climate change and of other environmental problems. Nevertheless, many parts of the media persist in treating conflicting views far too uncritically, obscuring and confusing the broad consensus that exists among researchers in these fields and the nature of the unresolved issues.

CHAPTER 15

The Second Half of the
Twentieth Century

On the most elemental level developments of the past half century can be described simply as having transformed a data-poor field to a data-rich one. This occurred through remarkable developments in instrumentation and data acquisition and processing that have made remote sensing and/or numerical modeling the central modes of nearly all current atmospheric research. Most important, the resulting information has stimulated major scientific advances that in turn give direction to further technical development.

A catalog of achievements of the past half century, far too extensive and complex to summarize usefully here, would necessarily highlight the following themes. First, observations and numerical modeling have extended weather forecasts, opened climate projection to scientific study and practical application, and revealed the potential threat of human-induced erosion of stratospheric ozone. Second, air–sea interactions, although long recognized as dominant influences on the global energy system, in the past have been largely inaccessible to systematic study due to the great difference in the scales of atmospheric and oceanic phenomena and because ocean data have been lacking. Though the ocean is still inadequately observed, these interactions on many scales have been brought within the envelope of systematic scientific study through satellite and buoy observations coupled with numerical modeling of the ocean–atmosphere system. Third, isotopic analysis of ice and sediment cores have made it possible to read records of climate change extending hundreds of thousands of years into the past with exquisite detail and opened a research field of great potential.

These achievements did not occur simply because talented scientists and engineers applied their abilities to them. Equally essential have been the development of institutions and the cooperative efforts of scientific and government organizations in mounting major coordinated research programs. The NAS, UCAR–NCAR, the AMS, government agencies and their coordinating adjuncts, WMO, ICSU, the IPCC, and other organizations provided the network of sinews essential to the actions described here. Of course there are defects in operation of the present system. Changes needed to strengthen the current structure are identified in chapter 12 of my book on global change (Fleagle 1994), and the future no doubt will require further changes.

The decades since the 1940s have witnessed changes in applications of the atmospheric sciences as dramatic as those of the sciences themselves. As this memoir opened simple in situ observations of temperature, wind velocity, precipitation, etc., were applied directly to immediate decisions relating to agriculture, aviation, and public safety. Manipulation of data was limited for the most part to simple interpretations of observations and to statistical summaries of data. Now, highly sophisticated remote sensing and data processing systems are used to generate basic datasets far more extensive and complete than was possible a half century ago. These data are used in a variety of complex applications that include, in addition to detailed synoptic analysis of weather, exploration of the atmospheres of Mars and other planetary bodies, radiation received from the sun, ocean surface properties, extent of sea ice, and many other geophysical and ecological assessments.

In the important area of global environmental change applications have already impinged frontally on major policy decisions and have brought atmospheric scientists into the political arena far more directly than ever before. Those entering the atmospheric sciences today face a radically different set of opportunities and challenges than did those of us who entered the field in the 1940s.

Acknowledgments

I am especially grateful to the following colleagues who read an earlier draft of this memoir and have corrected errors and identified many points needing clarification or elaboration: William Aron, retired director, NOAA Alaska Fisheries Science Center; David Atlas, NASA Distinguished Visiting Scientist; Franklin Badgley, UW Professor Emeritus of Atmospheric Sciences; Joost Businger, UW Professor Emeritus of Atmospheric Sciences and Geophysics; James Fleming, Colby College Professor of Science, Technology and Society; Conway Leovy, UW Professor Emeritus of Atmospheric Sciences and Geophysics; Richard Reed, UW Professor Emeritus of Atmospheric Sciences; Edward Wenk, UW Professor Emeritus of Engineering and Public Affairs.

Glossary

AAAS	American Association for the Advancement of Science
AAF	Army Air Forces
AF	Air Force
AEC	Atomic Energy Commission
AGU	American Geophysical Union
AMPS	Assembly of Mathematics and Physical Sciences
AMS	American Meteorological Society
BoB	Bureau of the Budget
BOMEX	Barbados Oceanographic and Meteorological Experiment
BSRL	Boeing Scientific Research Laboratory
CAS	Committee on Atmospheric Sciences
CCM	Certified Consulting Meteorologists
CFCs	Chlorofluorocarbon gases
CIA	Central Intelligence Agency
COMET	Cooperative Program for Operations, Meteorological Education, and Training
COMPUP	Committee on Public Policy
COPEPP	Committee on Public Education and Public Policy
COSPAR	Committee on Space Research
CSAGI	Comité Spécial de l'Année Geophysique Internationale
DoE	Department of Energy
E and G	Evaluation and Goals
EPA	Environmental Protection Agency
EPRI	Electric Power Research Institute
ESSA	Environmental Science Services Administration
FAA	Federal Aviation Administration
FBI	Federal Bureau of Investigation
GARP	Global Atmospheric Research Programme
GATE	GARP Atlantic Tropical Experiment
GFDL	Geophysical Fluid Dynamics Laboratory
GRD	Geophysics Research Directorate
GSFC	Goddard Space Flight Center
IAMAP	International Association of Meteorology and Atmospheric Physics
IAMAS	International Association of Meteorology and Atmospheric Sciences
IAS	Institute for Advanced Study
ICAS	Interdepartmental Committee for Atmospheric Sciences
ICO	Interagency Committee on Oceanography
ICSU	International Council of Scientific Unions
IGBP	International Geosphere–Biosphere Programme
IGY	International Geophysical Year
IPCC	Intergovernmental Panel on Climate Change

IUGG	International Union of Geodesy and Geophysics
JISAO	Joint Institute for Study of the Atmosphere and Ocean
JOC	Joint Organizing Committee
MIT	Massachusetts Institute of Technology
NACOA	National Committee on Oceans and Atmospheres
NAS	National Academy of Sciences
NAS/CAS	National Academy of Sciences/Committee on Atmospheric Sciences
NASA	National Aeronautics and Space Administration
NASCO	National Academy of Sciences Committee on Oceanography
NATO	North Atlantic Treaty Organization
NCAR	National Center for Atmospheric Research
NHRE	National Hail Research Experiment
NOAA	National Oceanic and Atmospheric Administration
NRC	National Research Council
NSF	National Science Foundation
NYU	New York University
OMB	Office of Management and Budget
ONR	Office of Naval Research
OST	Office of Science and Technology
PMEL	Pacific Marine Environmental Laboratory
PSAC	President's Science Advisory Committee
SCEP	Study of Critical Environmental Problems
RANN	Research Applied to National Needs
SCOPE	Scientific Committee on Problems of the Environment
SDI	Strategic Defense Initiative
STAC	Scientific and Technical Activities Commission
STREX	Storm Transfer and Response Experiment
UCAR	University Corporation for Atmospheric Research
UCLA	University of California at Los Angeles
UNEP	United Nations Environmental Programme
UOP	UCAR Office of Programs
USSR	Union of Soviet Socialist Republics
UW	University of Washington
WCP	World Climate Programme
WCRP	World Climate Research Programme
WMO	World Meteorological Organization
WWW	World Weather Watch

UW Faculty, September 2000

Academic Faculty

Marcia B. Baker, Professor, Geophysics and Atmospheric Sciences

David S. Battisti, Professor, Atmospheric Sciences

Christopher S. Bretherton, Professor, Atmospheric Sciences and Applied Mathematics

Dale R. Durran, Professor, Atmospheric Sciences

Qiang Fu, Assistant Professor, Atmospheric Sciences

Gregory J. Hakim, Assistant Professor, Atmospheric Sciences

Dennis L. Hartmann, Professor, Atmospheric Sciences

Peter V. Hobbs, Professor, Atmospheric Sciences

James R. Holton, Professor and Chair, Atmospheric Sciences

Robert A. Houze, Professor, Atmospheric Sciences

Lyatt Jaegle, Assistant Professor, Atmospheric Sciences

Clifford F. Mass, Professor, Atmospheric Sciences

Peter B. Rhines, Professor, Oceanography and Atmospheric Sciences

Edward S. Sarachik, Professor, Atmospheric Sciences

John Michael Wallace, Professor, Atmospheric Sciences

Stephen G. Warren, Professor, Atmospheric Sciences and Geophysics

Research Faculty

Robert A. Brown, Research Professor, Atmospheric Sciences

David S. Covert, Research Professor, Atmospheric Sciences

Thomas C. Grenfell, Research Professor, Atmospheric Sciences

Dean A. Hegg, Research Professor, Atmospheric Sciences

Igor V. Kamenkovich, Research Assistant Professor, Atmospheric Sciences

John D. Locatelli, Research Associate Professor, Atmospheric Sciences

Gary A. Maykut, Research Professor, Atmospheric Sciences

Bradley F. Smull, Research Associate Professor, Atmospheric Sciences

James E. Tillman, Research Professor, Atmospheric Sciences

Sandra E. Yuter, Research Assistant Professor, Atmospheric Sciences

Emeritus Faculty

Franklin I. Badgley, Professor Emeritus, Atmospheric Sciences

Joost A. Businger, Professor Emeritus, Atmospheric Sciences

Robert J. Charlson, Professor Emeritus, Atmospheric Sciences

Robert G. Fleagle, Professor Emeritus, Atmospheric Sciences

Halstead Harrison, Associate Professor Emeritus, Atmospheric Sciences

Edward R. LaChappelle, Professor Emeritus, Atmospheric Sciences and Geophysics

Conway B. Leovy, Professor Emeritus, Atmospheric Sciences and Geophysics

Richard J. Reed, Professor Emeritus, Atmospheric Sciences

Norbert Untersteiner, Professor Emeritus, Atmospheric Sciences and Geophysics

References

Benedick, R. E., 1991: *Ozone Diplomacy: New Directions in Safeguarding the Planet.* Harvard University Press, 300 pp.

Bjerknes, V., J. Bjerknes, H. Solberg, and T. Bergeron, 1933: *Physikalische Hydrodynamic.* J. Springer, 799 pp.

Bolin, B., et al., Eds., 1979: *The Carbon Cycles.* COPE Rep. 13, John Wiley, 491 pp.

——, et al., Eds., 1981: *Carbon Cycle Modelling.* SCOPE Rep. 16, John Wiley, 390 pp.

Bonner, W. D., and J. Paegle, 1970: Diurnal variations in boundary layer winds over the south-central United States in summer. *Mon. Wea. Rev., 98,* 735–744.

Buajitti, K., and A. K. Blackadar, 1957: Theoretical studies of diurnal wind-structure variations in the planetary boundary layer. *Quart. J. Roy. Meteor. Soc., 83,* 486–500.

Byers, H. R., and R. R. Braham, Eds., 1949: The Thunderstorm. Report of the Thunderstorm Project, U.S. Dept. of Commerce, Washington, DC, 287 pp.

Charney, J. G., 1947: The dynamics of long waves in a baroclinic westerly current. *J. Meteor., 4,* 135–165.

Committee on Atmospheric Sciences, National Research Council, 1971: *The Atmospheric Sciences and Man's Needs: Priorities for the Future.* National Academy of Sciences, 88 pp.

——, 1975: *Atmospheric Chemistry: Problems and Scope.* National Academy of Sciences, 130 pp.

——, 1977: *The Atmospheric Sciences: Problems and Applications.* National Academy of Sciences, 124 pp.

Crutzen, P. J., and J. W. Birks, 1982: The atmosphere after a nuclear war: Twilight at noon. *Ambio, XI,* 114–125.

Eady, E. T., 1949: Long waves and cyclone waves. *Tellus, 1,* 33–52.

Fleagle, R. G., 1945: The field of vertical motion in selected weather situations. *Trans. Amer. Geophys. Union, 26,* 359–363.

——, 1947: The fields of temperature, pressure, and three-dimensional motion in selected weather situations. *J. Meteor., 4,* 165–185.

——, 1948: Quantitative analysis of factors influencing pressure change. *J. Meteor., 5,* 281–292.

——, 1953: A theory of fog formation. *J. Mar. Res., 16,* 43–50.

——, 1957: On the dynamics of the general circulation. *Quart. J. Roy. Meteor. Soc., 83,* 1–20.

——, 1968: *Weather Modification: Science and Public Policy.* University of Washington Press, 147 pp.

——, 1970: Summary of Symposiom on "Early Results from BOMEX." *Bull. Amer. Meteor. Soc., 51,* 319–325.

——, 1978: Organization of the climate program. *Science,* **200,** p 9.

——, 1984: NOAA's RTD budget. *Science,* **225,** 365 pp.

——, 1986: NOAA's role and the national interest. *Sci. Technol. Hum. Values,* **11,** 51–62.

——, 1987: The case for a new NOAA charter. *Bull. Amer. Meteor. Soc.,* **68,** 1417–1423.

——, 1991: Policy implications of global warming for the Pacific Northwest. *Northwest Environ. J.,* **7,** 329–343.

——, 1992a: The U.S. response to global change: Analysis and appraisal. *Climatic Change,* **20,** 57–81.

——, 1992b: From the International Geophysical Year to global change. *Rev. Geophys.,* **30,** 305–313.

——, 1994: *Global Environmental Change: Interactions of Science, Policy, and Politics in the United States.* Praeger Publishers, 243 pp.

——, 1998: Northwest climate is changing: Loss of snowpack justifies efforts to replace fossil fuels with renewable sources of energy. *Seattle Times,* 12 April 1998.

——, and J. A. Businger, 1963: *An Introduction to Atmospheric Physics.* Academic Press, 346 pp.

——, and J. A. Businger, 1980: *An Introduction to Atmospheric Physics.* 2d ed. Academic Press, 432 pp.

——, and E. L. Wolff, 1979: The creation and uses of public policy: An AMS Symposium on Atmospheric Science Policy, 24–25 May 1978, Boston, Mass. *Bull. Amer. Meteor. Soc.,* **60,** 638–648.

——, J. A. Crutchfield, R. W. Johnson, and M. F. Abdo, 1974: *Weather Modification in the Public Interest.* Amer. Meteor. Soc., 88 pp.

Holton, J. R., 1967: Diurnal boundary layer wind oscillation above sloping terrain. *Tellus,* **19,** 199–205.

House Committee on the Budget, 1978: *Hearings on Task Force on Community and Physical Resources.* 95th Cong., 2d sess., 14 March 1978, 27–41.

Humphreys, W. J., 1929: *Physics of the Air.* 2d ed. McGraw Hill, 654 pp.

Interdepartmental Committee for Atmospheric Sciences, 1964: The Federal Atmospheric Sciences Program for FY 1966. Federal Council for Science and Technology Rep. 9, Washington, DC, 55 pp.

Jacobs, W. W., 1942: On the energy exchange between sea and atmosphere. *J. Mar. Res.,* **5,** 37–66.

Kellogg, W. M., D. Atlas, D. S. Johnson, R. J. Reed, and K.C. Spengler, 1974: Visit to the People's Republic of China: A report from the A.M.S. delegation. *Bull. Amer. Meteor. Soc.,* **55,** 1291–1330.

Kennedy, J. F., 1962: *Public Papers of the President of the United States.* U.S. Govt. Printing Office, 622 pp.

London, J., and G. F. White, Eds., 1984: *The Environmental Effects of Nuclear War: AAAS Selected Symposium 98.* Westview Press, 203 pp.

NAS/CAS, 1962: The atmospheric sciences, 1961B1971. NAS-NRC Publication 946, Washington, DC, Vol. 1, 89 pp; Vol. 2, 99 pp; Vol. 3, 36 pp.

———, 1966: The feasibility of a global observational and analysis experiment. Panel on International Meteorological Cooperation, Publication 1290, NAS–NRC, Washington, DC, 172 pp.

———, 1971: *The Atmospheric Sciences and Man's Needs: Priorities for the Future.* National Academy of Sciences, 88 pp.

NAS–NRC, 1970: Educational Implications of the Global Atmospheric Research Program. *Bull. Amer. Meteor. Soc., 51,* 327–333.

Nebeker, F., 1995: *Calculating the Weather.* Academic Press, 255 pp.

Petterssen, S., 2001: *Weathering the Storm: Sverre Petterssen, the D-Day Forecast and the Rise of Modern Meteorology.* J. Fleming, Ed., Amer. Meteor. Soc., 329 pp.

Phillips, N. A., 1956: The general circulation of the atmosphere: A numerical experiment. *Quart. J. Roy. Meteor. Soc., 82,* 123–164.

———, 1998: Carl-Gustaf Rossby: His times, personality, and actions. *Bull. Amer. Meteor. Soc., 79,* 1097–1112.

Physics Survey Committee/National Research Council, 1973: *Physics in Perspective.* Vol. 2. *Part B: The Interfaces,* National Academy of Sciences, 1465 pp.

PSAC Panel on Oceanography, 1966. *Effective Use of the Sea.* U.S. Government Printing Office, 144 pp.

Rader, M., 1969: *False Witness.* University of Washington Press, 209 pp.

Sanders, J., 1979: *Cold War on the Campus: Academic Freedom at the University of Washington, 1946–1964.* University of Washington Press, 243 pp.

Stewart, G. R., 1941: *Storm.* Random House, 349 pp.

Study of Critical Environmental Problems (SCEP), 1970: *Man's Impact on the Global Environment; Assessment and Recommendations for Action; Report.* MIT Press, 319 pp.

Turco, R. P., O. B. Toon, T. Ackerman, J. B. Pollack, and C. Sagan, 1983: Nuclear winter: Global consequences of multiple nuclear explosions. *Science, 222,* 1283–1292.

U.S. Committee for GARP, 1969: Plan for U.S. Participation in the Global Atmospheric Research Program. NAS-NRC, Washington, DC, 79 pp.

———, 1975: Understanding Climate Change: A Program for Action. NAS–NRC, Washington, DC, 239 pp.

U.S. Senate Committee on Commerce, Science, and Transportation, 1983: *Hearings before the Committee on Commerce, Science, and Transportation.* 98th Cong., 1st sess., 14 March 1983, 222–230.

———, 1987: *Hearing on NOAA Authorization for Atmospheric and Satellite Programs.* 100th Cong., 1st sess., 19 May 1987, 40–51.

U.S. Senate Committee on Foreign Relations, Subcommittee on Oceans and International Environment, 1972: *Prohibiting Military Weather Modifications.* 92d Cong., 2d sess., S. Res. 291, 162 pp.

Wenk, Edward, Jr., 1972: *The Politics of the Ocean.* University of Washington Press, 87–154.

Index